History, Philosophy and Theory of the Life Sciences

Volume 25

More information about this series at https://link.springer.com/bookseries/8916

Kolja Ehrenstein

Causal Pluralism in the Life Sciences

A Journey Along the Frontiers of Conceptual Plurality

Springer

Kolja Ehrenstein
Köln, Germany

ISSN 2211-1948 ISSN 2211-1956 (electronic)
History, Philosophy and Theory of the Life Sciences
ISBN 978-3-030-87941-9 ISBN 978-3-030-87942-6 (eBook)
https://doi.org/10.1007/978-3-030-87942-6

This Springer imprint is published by the registered company Springer Nature Switzerland AG.
The registered company address is: Gewerbestrasse 11, 6330 Cham, Switzerland

In gratitude dedicated to my grandfather, Kurt Gerntke, and my godmother, Gertrud Ehrenstein, who both passed away while I was preparing the material for this book

Preface

The material in this book is derived from my doctoral dissertation, which I submitted in September 2018 to the Faculty of Arts and Humanities at the University of Cologne (Philosophische Fakultät der Universität zu Köln). The original title of that work was 'Causal Pluralism in the Philosophy of Biology'. This work was supervised by Andreas Hüttemann (Department of Philosophy). The other members of the examination commission were Thomas Grundmann (Philosophy), Siegfried Roth (Biology), and Christine Chwaszcza (Philosophy) who chaired my examination (defensio).

I will briefly explain how the title was originally chosen. When I started the project, I did not yet have a firm idea of what 'causal pluralism' could be. It seemed to be clear that it was a popular yet at the same time vague position. An area where causal pluralism gets much attention is in the philosophy of biology. My plan was to bring this confluence into sharper focus and to decide whether it is reasonable to adopt a position of causal pluralism. This meant that I had to choose between various versions of causal pluralism (Sect. 1.2) in order to get my project started. I eventually settled on the idea of treating causal pluralism from a semantic perspective. The subtitle now points the reader towards this more specific approach.

The semantic approach is quite unusual when one considers how the word 'pluralism' is used in the philosophy of science, where it often indicates that someone is opposing reduction or rejecting a certain dogmatism. In order to avoid raising false expectations, I have now chosen a more specific title. My approach is to think of causal pluralism in terms of conceptual subsumption (Chap. 3), and my aim is to explore where such pluralism can be found and how much of such causal pluralism is reasonable. The conclusion I reach in this book is that there are few benefits to conceptual pluralism and that many authors have a tendency to overestimate the perceived advantages of conceptual plurality. I particularly caution philosophers against making new conceptual suggestions to biologists without a thorough analysis of whether such concepts actually contribute to solving the sorts of problems that are debated in the life sciences.

One might imagine that a larger conceptual apparatus should do no harm and that it is always better to use a more specific term in place of a more general one.

This, however, is not the case, and I shall argue that discourse in the life sciences has indeed suffered from the use of irrelevant and even misleading causal concepts. The problem as I see it is that authors can too easily draw attention to themselves by coining new—but essentially hollow—concepts.

This book is aimed at philosophers of science, and at biologists who take an interest in the philosophical analysis of the life sciences. It should be possible to read this book from either of those two perspectives. I have included introductory material for anyone who is not familiar with causal theory (Chap. 2), and while writing, I have also kept in mind that the text should be readable without a special training in biology (as long as someone has a general knowledge of the natural sciences). My hope is that this book will also be accessible to other readers, although this can only be decided on an individual basis by those who have it in their hands.

The 'story' and 'moral' of this book are after all universal, in the sense that I think similar books could be written about debates in any number of disciplines: I am critical of a conceptual distinction becoming popular despite its lack of a sound theoretical account (Chap. 4); I argue that other quarrels could have been avoided with a more careful study of primary sources (Chap. 5); I point out that a theory may be beautiful without fitting the concrete case to which it is applied (Chap. 6); and I warn against letting other cases tempt us into adopting strong versions of such theories (Chap. 7). Arguments with similar structures can be made about all sorts of issues.

Yet this book is still different from a typical philosophical essay. You may find it easier to follow the argument in this book if you abandon the expectation that I am going to refute my opponents by disproving their positions. This is not my primary strategy, since pluralism is a very flexible and hard-to-grasp doctrine.

'Flexible doctrine' may appear to be a contradiction in terms. We tend to think of a doctrine as a solidified, hardened belief. This suggests that a doctrine is a clearly defined idea or position. However, if this were the case, indoctrination would be less of a problem, since clearly defined ideas are those ideas that are easily criticised and discarded if they are nonsense. Positions that are vague, opaque or flexible are much harder to criticise, and these positions, therefore, have much more potential to turn into a doctrine.

Since pluralism is not clearly defined, I am not following the standard procedure of seeking contradictions in my opponent's position. I will occasionally draw attention to a contradiction, but this is only a peripheral aspect of my overall argument. My principle strategy is to argue that *pluralism does not deliver on its promises*.

Sometimes, I will also argue that what appears to be pluralism is unlikely to be a persuasive instance of pluralism. However, bear in mind that, although I do try to get the idea of causal pluralism into sharper focus, it remains an obvious evasive strategy to claim that that I have identified the wrong sort of pluralism. This shows how important it is to pose the question of what causal pluralism actually accomplishes.

The work on the material in this book began in 2011 when I moved to Cologne for my graduate studies, having previously studied biology and philosophy at the

University of Münster, where I graduated in 2008. I liked studying philosophy in Münster very much. The Philosophy Department had good teachers and a cosy (though somewhat neglected) seminar building in the city centre. It was also in Münster where I met Andreas Hüttemann, who first supervised my state examination thesis in Münster and later my doctoral dissertation in Cologne.

I may perhaps share an anecdote that (partially) explains why I chose Andreas as a supervisor. Although he would not remember this, one of *my* key interactions with him was a discussion in a seminar. The subject matter is not relevant here (I think it was something in the philosophy of mind). What happened was that I first answered a question to the best of my knowledge until Andreas kept persisting, asking me to clarify just one further point. That was when I had to admit that I did not understand what the author we were discussing intended to say. To my relief, Andreas went on something of a rant, not against me, but against that author, explaining to us students why it was impossible to understand what that author was saying. This was the first moment when I seriously considered choosing that professor as a supervisor.

I am glad that my intuition did not deceive me and that Andreas gave me great advice over the years—not just subject-specific knowledge but also advice that made it easier to get through my dissertation project. When I got sidetracked by a problem, he would ask whether that detail was really crucial to my argument and whether I could not simply drop that thought. When there was something in my manuscript that was crucial but could easily be overlooked, he would suggest repeating it to make it stand out. His attitude was that this was not a beauty contest and that I could edit the material later for publication. I thank Andreas for his guidance and support over the years!

I apologise to anybody who found it a bit dull to work through my initial dissertation manuscript. However, the repetitions paid off on the day of my defensio. I had to answer many thorough questions, yet none arose from something that was overlooked because I had mentioned it only briefly in my manuscript. I thank Thomas Grundmann and Siegfried Roth for their questions on that day, and I extend a further thank-you to Christine Chwaszcza who stayed a short time after my defensio to help us celebrate. My sister Ulrike, who had set up a camping table with some drinks and snacks for us, later remarked what pleasant company Professor Chwaszcza was.

But, let me return to the writing process. Most of the discussion of the drafts for my dissertation project took place in the colloquium, i.e. the weekly meeting of Andreas's staff, graduate students and occasional guests from outside. Over the years, it has been difficult to keep track of all the input that I received at those meetings. I am grateful for the many suggestions, even if I cannot attribute all of them personally. There are two sections, Sects. 3.1 and 7.1, in which I have placed footnotes to acknowledge the individual support that brought those sections into the specific forms in which they still appear in this work.

At the time I arrived in Cologne, there also existed a research group for causation, laws, dispositions and explanation. I was never a formal member of that group, but I had the benefit of attending the conferences that it organised. I was also invited to present talks at two of those conferences. I thank Marie Kaiser for those invitations.

Few of the ideas that I presented at those conferences made it into this manuscript, but the discussions I had were important.

The argument in this book is largely the same as that in my doctoral dissertation, yet there are quite a few adjustments that I made afterwards. I did streamline and often clarify my arguments, and I also added a completely new chapter on permissive and instructive causes (Chap. 5). The debate that I address in that chapter did not break out until after the date I submitted my dissertation, which is why I did not discuss it there. I am grateful to the two anonymous reviewers who supplied constructive criticism that helped to create this improved version of my original work. I thank the editors of the series for seeing the potential in my original draft, particularly Thomas Reydon with whom I corresponded. I also thank Svetlana Kleiner at Springer for the detailed answers she gave me when I had questions about the manuscript preparation process.

The text in this book was proofread by Dean Jennings who also alerted me to many ambiguous expressions or passages that were susceptible to misinterpretation. I thank Dean for the careful consideration of my original draft and the advice he gave me. Some decisions I could never have taken myself, such as applying a plural-*s* to a mass noun under very specific circumstances. This can only be done by a native speaker. I recommend his scholarly proofreading service for any project that depends on careful language choices.

The completion of this work would have been impossible without my parents. I thank them for their constant support. I am also grateful that I live in a country with affordable higher education, although family and socio-economic background are still important factors. Following through on a dissertation project is much easier if you are surrounded by people who value education.

I thank my brother Tilman for reading some very early drafts and helping me to sort my ideas. When I had to re-edit the manuscript for publication, I received some practical assistance from my dear school friend Ronny Eckardt. I also thank him for sharing some of his medical knowledge, which proved to be most useful. I am grateful to all those who kept me good company. Some of you took an honest interest in the strange thoughts that occupied my mind. It can be challenging to give useful answers . . . when you are in a pub in northern France being questioned about possible-world semantics. My final token of gratitude goes to Ulrike, for preparing that camping table with snacks and drinks.

And now that all this is over, I am looking forward to the end of this dreadful pandemic, so that we can once again meet up with each other in those places that have been closed all over the world. I invite you all—and particularly, *you* who have read through all these acknowledgements—to join me for a drink in Chap. 1!

Cologne, Germany
August 2021

Kolja Ehrenstein

Contents

Chapter 1
Conceptual Pluralism

Abstract This chapter introduces the problem of conceptual plurality. I argue that the best way to approach causal pluralism is to take a semantic approach, as opposed to an epistemic or metaphysical approach. This chapter gives a brief preliminary introduction to my account of conceptual causal pluralism and I explain why I reject the idea of proving or disproving the vague general position of causal pluralism.

Keywords Arity pluralism · Conceptual causal pluralism · Causal monism · Causal pluralism · Epistemological-methodological causal pluralism · Language · Metaphysical causal pluralism · Scepticism · Scientific realism

Everyone is familiar with the sport of squabbling over words. In my home town, you can walk into a pub and order a *beer*, and the bartender will tersely respond, 'You mean one *Warsteiner*, then?'—leaving you thinking to yourself, 'Why are you correcting me?—We both know that's the *only* beer sold in town.' A related but different situation might also cause a problem. You can walk into a shop, describe the item that you are seeking, and be offered something similar—but think to yourself, 'No, that simply won't be an acceptable replacement for the chair I accidentally broke last weekend.'

There are important asymmetries between the two stories: In the first scenario the bartender is simply forcing a word on you. In the second scenario no such arbitrary force is exerted. In the first scenario it doesn't matter what that beer is called. You won't cancel the order and the bartender won't cancel the delivery. In the second scenario the exact linguistic specification of the item is of utmost importance. If you cannot replace what is lost, your only remaining hope is that what is now gone won't be missed. In the first scenario you are colliding with language. In the second scenario you are colliding with reality.

Researchers in the life sciences seek that confrontation with reality. That is after all the point of any empirical science. This means that hammering home a brand message instead of a relevant claim can act disruptively in science.

How is one to tell apart these two cases? Demanding that the concepts have empirical content won't distinguish them. Telling one brand apart from another is

K. Ehrenstein, *Causal Pluralism in the Life Sciences*,
History, Philosophy and Theory of the Life Sciences 25,
https://doi.org/10.1007/978-3-030-87942-6_1

a perfectly empirical matter. There are clear truth conditions for testing whether a beer is Brand V or Brand W.

We also need to take into account the fact that people use innocent synonyms. Not every change or substitution of an expression is malicious. Sometimes you want to avoid repeating a word within too short an interval. In scientific literature it may be acceptable to repeat oneself more often than in a novel, but even in the scientific genre synonyms enrich language and are important for keeping a reader's attention.

If someone is over-specific in communication, this is also not necessarily due to malicious intent. This is how the name of the Czech city Pilsen became synonymous with 'lager beer' in the German language.

Perhaps one can argue that a principled account is not really needed. The educated recipient of a message will often recognise straight away when a particular phrase is used for dubious advertisement instead of plain communication.

Although there is probably no general method for sorting out manipulative from unpretentious messages, the problem is real. Some have argued that it is also an issue in the life sciences, where developmental biology is said to have taken on an outsider status after the Modern Synthesis (i.e. the formation of the integrated or synthesised theory of evolution), which persisted up until the time it experienced a renaissance in the field of evolutionary developmental biology—evo-devo research, for short.

Some have blamed one person in particular for the chequered history of developmental biology. In the common narrative, evolutionary biologist Ernst Mayr is considered to have—arbitrarily—pushed the idea of there being ultimate and proximate causes, thus rendering developmental biology invisible even long after the synthesis had taken place.

I don't share this simplified view. The extensive historical analysis by Ron Amundson (2005) shows that there is much more to the story than an idea merely spiralling out of control. Developmental biologists have in my opinion also ineptly attacked that idea and clumsily defended their own position.

Unlike Amundson, I shall not conduct a historical analysis in this book. My plan is instead to throw the wider philosophical theory of causation at Mayr's idea—and, while I am at it, also to discuss the idea of permissive and instructive causes, and more general philosophical ideas. For everyone knows it is philosophers who are the ones that trade in the truly manipulative and addictive ideas.

This book examines the way that similar ideas reappear in different theories of causation. By placing Mayr's thesis (that one has to distinguish ultimate from proximate causes) in the context of other philosophical ideas, I intend to dismantle even the remnants of it which have been left behind after previous attacks.

The other parts of this book can be read as a guidebook to the limits of conceptual causal pluralism: What philosophical idea can be chosen to interpret the notion of there being permissive and instructive causes? Is the philosophical distinction between production and dependence interesting from a biologist's perspective? And should one opt for a specific philosophical theory (or theories) when considering the fact that virtually every causal mechanism in biology involves causation by disconnection? I have also included an idea of how to bring order to the notion of causal pluralism.

1.1 A Salesperson's Paradise

Philosophers have been selling a number of theories of causation over recent centuries. In particular, the period from the mid-twentieth century up until the present day has seen a proliferation of competing theories. For a long time before that, however, the discourse on causality was dominated by David Hume's regularity theory of causation (Sect. 2.1), which combined a metaphysical analysis of causation (as regular succession) with a semantic analysis of 'causation' (as necessary connection). Although John Stuart Mill improved on Hume's original account, it turned out that a regularity theory couldn't keep pace with scientific developments.

In the late nineteenth and early twentieth centuries, this led firstly to a rejection of the notion of cause. Ernst Mach (1883/2012) and later Bertrand Russell (1912–1913) criticised the thesis that there are regularities in nature. They argued that—strictly speaking—nothing repeats itself in nature, and that science didn't describe nature in terms of regular succession but in terms of mathematical functions. From this they concluded that the notion of cause was vague and unscientific.

A shift occurred when Hans Reichenbach (1930) began considering the possibility of probabilistic causality. This is significant because Reichenbach didn't follow Mach and Russell in rejecting the notion of cause as unscientific. He adopted a new view which left behind both of Hume's analyses, semantic and metaphysical. The objection that there are no regularities in nature doesn't apply any longer to a view such as Reichenbach's.

Later G. E. M. Anscombe (1971/1993) even argued that Hume's semantic analysis had been flawed all along. The folk notion of causality had always been something else. Here I am uncertain whether such a position can escape the charge of being vague and unscientific. However, the point is that in the twentieth century all sorts of views became acceptable. Today we are living in the age of causal pluralism, not just because our minds have been liberated from thinking that causation means necessary connection or regular succession, but also out of despair.

The general sentiment which seems to be shared by a variety of authors is that it is very difficult to come up with a *single* theory of causation (Hitchcock, 2003; Cartwright, 2004; Hall, 2004; Campaner & Galavotti, 2007; Glennan, 2009; Psillos, 2010). This points to a notion that causal pluralism essentially means that there is no universally accepted theory of causation.

On top of this there also appears to be confusion about the term 'pluralism' itself. Christopher Hitchcock (2007) discusses some nine different meanings that 'causal pluralism' may have.

In the current situation one is spoiled for choice. You don't like Brand V? You are right, Brand W suits you much better. It has twice as many strokes in its name and is easily customisable to your needs. And about that little 'accident' of yours—I've got you covered there, too! Here is an amazing product. It ships with an empty box and a watertight argument written on it. That should let you off the hook! It's called the 'quasi-chair'.

1.2 Singling Out a Relevant Notion of Causal Pluralism

The contemporary market has much to offer, so I would suggest not getting lost in the details but instead following a useful threefold distinction drawn by Leen De Vreese (2010) between epistemological-methodological, metaphysical, and conceptual versions of causal pluralism. I believe that this is a very natural systematisation, which is also reminiscent of Mackie's 'factual, conceptual, and epistemic' analyses of causation (Mackie, 1980, p. ix).

Although I may not interpret these three versions of causal pluralism in the same way as De Vreese, I shall use her three labels for my sorting baskets anyway, and supply comment on the content which I throw into each of them. I begin with:

1. **epistemological-methodological causal pluralism**: the thesis that there are many epistemological or methodological means of uncovering causal relations.

Epistemological-methodological causal pluralism appears to be a natural assumption. Christopher Hitchcock argues in favour of this view: pharmaceutical studies rely on controlled experiments, social studies infer causal relations from observation and statistical analysis, forensic studies use fingerprints as clues to causation, and so on (Hitchcock, 2007, p. 204).

If one wanted to defend this idea further, an argument could be made from a realist perspective. Wesley C. Salmon defended scientific realism with an argument about Avogadro's number. He argued that the fact that researchers determined Avogadro's number through various *independent* and differently designed experiments strongly suggests that there are indeed entities to be counted, i.e. that there are atoms or molecules, although this still leaves much about the specific reality of atoms unanswered (Salmon, 1984, pp. 214 ff.).

Salmon's argument rests on the principle of the common cause (Sect. 2.3), which stipulates that improbable correlations have to be accounted for by a common cause. The fact that there are $6.022 * 10^{23}$ particles in a mole would be the common cause of the correlation of the results in the various experiments conducted for finding Avogadro's number.

To apply that argument to another example, the correlation between intake of ethylene glycol and sickness is accounted for by a common cause, in this case the fact that ethylene glycol gets metabolised by the alcohol dehydrogenase to glycoaldehyde, which then gets further metabolised, eventually causing the failure of various physiological processes (Battistella, 2002).

The plurality found here also applies to different experiments, which were conducted in order to conclude how ethylene glycol and the alcohol dehydrogenase interact. Uncovering how the alcohol dehydrogenase metabolises ethylene glycol requires many analyses, stretching from observations of correlations, to experiments uncovering the molecular structure of the substances, and to accounts of how exactly the respective covalent bonds in ethylene glycol are rearranged during an interaction with the alcohol dehydrogenase. Nevertheless, you may note a difference at this point: those experiments also support each other but more in the sense of *closing different gaps* in the causal explanation, each of them thus complementing instead of reaffirming the same exact result of another experiment. One can probably argue that researchers discovered those causal facts about ethylene glycol poisoning without having or needing a metaphysical theory of causation.

Hitchcock argues that one can extend epistemological-methodological causal pluralism to *causal theories*, too. Most theories of causation think of causation in terms of probability raising, counterfactual dependence, energy-momentum transfer, or something of that sort. But Hitchcock suggests that one could also hold the view that there is no need to identify causation with probability raising or anything else that has been taken as a criterion for causation. In that case, causal theories would merely help to uncover causation (whatever that may be) in different circumstances, and one could have different theories in different circumstances (Hitchcock, 2007, p. 205). Campaner and Galavotti (2007) take a similar view.

Although I find pluralism about scientific research methods extremely plausible, I am not sure whether such a pluralism about theories of causation is equally compelling. An obvious problem is that causal theories tend to make ontological assumptions. Regularity theory claims—falsely—that causation *is* regular succession (Sect. 2.1). According to Salmon's process theory, the world is full of causal processes capable of transmitting marks or (in the updated version) conserved quantities (Sect. 2.5). Counterfactual theory is a *conceptual* analysis that doesn't care much about ontology, except that it has to be an anti-realist ontology (Sect. 2.2).

Hitchcock suggests, of course, that one would have to abstract from the ontology. Although I agree (in principle) with that idea, and I will, when needed, apply a methodological agnosticism about the metaphysics of causation (as in Chap. 3), it would be misleading not to disclose from the beginning that I will discard certain causal theories later on in this book. To be consistent, I cannot advocate an unqualified version of this sort of epistemological-methodological causal pluralism. Why am I taking this stance, and in what sense am I restricting the notion of epistemological-methodological causal pluralism?

In my opinion, it is important that causal theories are capable of communicating with science—in particular with the natural sciences. Hume's theory, for instance, is an interesting (although probably outdated) psychological theory of how human beings form an idea of causality. His regularity hypothesis is nowadays universally rejected, despite Mackie's attempt to rescue the analysis in the form of complex regularities (Sect. 2.1). What remains is Hume's empiricism, which is a good thing, and an anti-realist metaphysics of causation, which is—to say the least—questionable. For me, Humean empiricism and Humean metaphysics are *not* inseparable. I take the view that Humean empiricism is compatible with a realist outlook on causality, as advocated by Salmon (2002/2018) or Hüttemann (2013).

Humean metaphysics survives today in the form of the counterfactual theory of causation (Sect. 2.2). Although I lean towards a realist position, counterfactual theory's anti-realism about causation is no reason for me to reject it. What concerns me is that counterfactual theory is joined to an account of causal explanation which is unsuited for biology (Sect. 7.6). I don't see this as an insurmountable difficulty, and I would welcome adjustments to that account of causal explanation. It appears, however, that proponents of counterfactual theory argue for their implausible view of causal explanation as a consequence of their metaphysical convictions. I don't want to read too much into their views, but, if this is true, their modus ponens for that far too liberal account of causal explanation is my modus tollens against counterfactual theory.

My acceptance of epistemological-methodological causal pluralism is limited to the idea that more than one theory of causation can act as a heuristic. Yet I don't think that every theory of causation is equally plausible as an ontic account. However, as long as theories converge towards a plausible ontic view, more than one theory can act as a useful heuristic. In my opinion, the best choice of ontic account is currently Andreas Hüttemann's (2013) disposition-based process theory, which also appears to be suitable as an account of causation in biology (Sects. 2.6 and 7.3). This theory postulates that causes are actual disturbances of the default behaviour of systems. An advantage of Hüttemann's theory is that it avoids the embarrassment of 'quasi-causation' to which traditional process theories are committed (Sect. 2.5).

As I have declared a preference for a specific metaphysical view of causation, I will not embrace any strong commitment to:

2. **metaphysical causal pluralism**: the thesis that there is more than one kind of causation.

But it is not just me backing a particular view of causation that gets in the way of debating this sort of pluralism. There is also the fact that the question of realism usually gets more attention. If Humean philosophers argue—strictly speaking—that there is no causation, then there is little left about which they could be pluralists. The only way the Humean could be a metaphysical causal pluralist would be to be a pluralist about the reduction base of causation, i.e. the Humean mosaic (Sect. 2.2). I have, however, no concrete idea how that could be sustained as a position.

This suggests that a prerequisite for metaphysical pluralism is realism. One could perhaps interpret some of the literature on mechanisms in biology as promoting metaphysical causal pluralism (Machamer et al., 2000). However, this version of causal pluralism is probably not the most important, since this is a rather narrow notion of pluralism which only captures a small part of the debate.

I suggest that more of the debate on causal pluralism becomes intelligible through the lens of:

3. **plurality of causal concepts (naive account)**: the thesis that there are many concepts of cause and effect.

Note that this is not yet my final thesis, but only the initial direction which I am taking from this point onwards. The obvious problem with this version of causal pluralism is that it tends to trivialise the issue. It is far too easy to introduce a new word, and even to provide empirical criteria for when to use such a word. The general direction is, nevertheless, the right one, if one considers examples of conceptual causal pluralism.

Hitchcock believes that there are several theoretically interrelated concepts of cause. He calls this 'intramural pluralism' and suggests that one should distinguish between concepts such as 'positive' and 'negative cause', or 'net' and 'component effect' (Hitchcock, 2007, pp. 207 ff.). These distinctions are supposed to be consistent with metaphysical causal monism. Conceptual causal pluralism need not imply metaphysical causal pluralism. Hitchcock claims that—metaphysically speaking—causation may still be one sort of 'thing', e.g. a relation of manipulability (Hitchcock, 2007, p. 210).

Another sort of conceptual causal pluralism is the two-concept view proposed by Ned Hall (2004), who argues that there are two concepts of cause that do not have much in common. He distinguishes 'dependence' from 'production'. These concepts are subject to two largely unrelated analyses of causation. How this sort of conceptual causal pluralism is related to metaphysical causal pluralism is not entirely clear. Hall admits it could be said that his argument merely shows 'that there are two kinds of causation' (but not two concepts of cause), and goes on to say that he has no firm idea of what the difference between concepts and kinds should be (Hall, 2004, p. 255). Assuming that there is a difference between kinds and concepts, one could say that Hall's account exemplifies the contrary case of 'intramural pluralism'. By an intramural pluralism account, a plurality of concepts is compatible with a metaphysically monist account of causation. But if one considers Hall's account, or his objection to his account, there could also be the reverse: a metaphysical plurality without a conceptual plurality.[1]

Although it is conceivable that conceptual monism may mask metaphysical pluralism, one may argue that this is a rather remote possibility. The fact that issues of metaphysics can be clearly articulated suggests that those issues are usually also reflected in the causal concepts applied. Raising the possibility of metaphysics

[1] See also Godfrey-Smith (2009, pp. 327 f.) for a discussion of such possibilities.

getting masked is perhaps better seen as a variant of Anscombe's critique of Hume's analysis. If, as Anscombe supposes, any analysis of the concept of cause is so abstract that it becomes virtually meaningless, then an underlying metaphysical plurality may be masked by the analysis. However, if it is also true, as supposed by Anscombe, that one can substitute 'cause' with more specific causal verbs in every instance, than such a masking should be of little consequence. All one would have to do would be to use those specific causal verbs, of which Anscombe gives examples such as 'scrape, push, wet, carry, eat, burn, knock over, keep off, squash' and so on (Anscombe, 1971/1993, p. 93 italics removed). In that case nothing is lost by analysing the general notion of cause (and nothing gained if Anscombe should be right).

Whatever the potential loss of metaphysical issues may be, I suggest that the best targets for an analysis of causal pluralism are the causal concepts. For one thing, there are at least some authors such as Hitchcock and Hall who have explicitly endorsed some form of conceptual causal pluralism. In addition, one may argue that analysing conceptual plurality leaves Humeanism on the table (although I prefer a realist approach). Finally, one may argue that metaphysical convictions or ideas about the epistemology of causation are still likely to be reflected in the causal concepts articulated by the various theories of causation—hence, even those two issues are not completely lost, since they intersect with the idea of conceptual plurality. What remains is the task of turning the naive idea of conceptual causal pluralism into a less trivial idea than stated above.

1.3 A Brief Outline of Conceptual Causal Pluralism

A common theory of concepts is that they are—loosely speaking—predicates (Frege, 1986). This means that one can try identifying a causal concept by extracting a predicate from a causal statement. Take, for instance, the judgement:

> The event that lightning occurred is a cause of the event that forest fire occurred.

The predicate in this sentence or causal judgement is:

. . . is a cause of . . .

As you can see, this predicate has two blanks. One is supposed to be completed with a term denoting a cause, the other with a term denoting an effect. The predicate connects two terms. Or, to put it into fancy language, the predicate has the degree 2.

An advantage of this approach to causal concepts is that one may now deal with them in terms of first-order logic. One may, for instance, assign a set of binary relations to the predicate (as the extension of that predicate):

C^2_4: {< ①, ② > | ① is a cause of ②}

The expression on the left-hand side of the colon is a (to some extent arbitrarily chosen) symbol for the predicate in first-order logic. The expression on the right-hand side of the colon tells us which set of relations belongs to the predicate.

This approach to concepts of cause and effect is in principle neutral to the question of whether Humeanism is true. One who is anti-realist about causation, i.e. the Humean, would argue that there is nothing *more* to causation than being a certain relation. The causal realist, by contrast, could argue that ① *necessitates* ②. (Modern realist approaches take a more sophisticated view.) Yet both the anti-realist and realist would agree on the fact that there is a set of relations called 'causal relations'. In this book I shall rely on this minimal consensus.

However, the interesting question is of course how to overcome the trivial idea that conceptual causal pluralism is plurality of causal concepts. This notion of pluralism is too simple since it is always possible to take a subset of the extension of C_4^2 and put a new label on it, e.g. 'C_{772}^2'. If a theory stipulates many such subsets of C_4^2, would that already constitute conceptual causal pluralism in a meaningful sense?

My view is that one should avoid that conclusion. I will, therefore, suggest adopting a different notion of conceptual causal monism and pluralism:

1. **Conceptual causal monism**: a theory of causation that claims to analyse a master concept of cause to which all other causal concepts are subsumable.
2. **Conceptual causal pluralism**: a theory of causation that does not claim to analyse a master concept of cause to which all other causal concepts are subsumable.

From this perspective the number of concepts is no longer a criterion. A monist theory can postulate many causal concepts as long as they are subsumable to the master concept. A pluralist theory need not postulate a high number of concepts as long as it refrains from analysing a master concept to which all other concepts are subsumable.

I shall argue in Chap. 3 that this notion of conceptual causal pluralism is practical, and that it sorts most accounts of causation into their intuitively right categories. An exception is Hitchcock's 'intramural pluralism', which, I shall argue, is monism rather than pluralism.

Another interesting aspect of Hitchcock's idea of 'intramural pluralism' is that it only incompletely fits my account of conceptual causal monism or pluralism. I shall suggest that this calls for an extension of the account of conceptual causal pluralism.

The odd thing is that there can be causal concepts with different arity. This means, for instance, that there is a pair of concepts (belonging to each other) where one concept or predicate has the degree 2, while the other concept or predicate has the degree 3. In order to accommodate this I postulate that there is also arity pluralism:

3. **Arity pluralism of causal concepts**: a theory of causation that proposes at least two concepts of cause differing in arity.

Although I think that this is a true second dimension of conceptual causal pluralism, a cautious remark seems appropriate. The question of the arity of a causal concept is to a certain degree open to interpretation. The reconstruction of a concept's arity is not completely arbitrary, but one has to be careful not to apply double standards. I shall discuss this in Sect. 3.1 and argue that a coherent interpretation policy is needed.

1.4 Putting Pluralist Ideas to the Test

As indicated at the beginning of this chapter, the fact of conceptual plurality in itself is nothing to be worried about. There are many legitimate and often innocent reasons for having a multitude of expressions. Problems may arise, however, when the use of certain words gets arbitrarily promoted. Whether this is what happened when Mayr pushed the idea of ultimate and proximate causes will be one question. Another overarching question will be whether philosophers have been trying to sell useless causal concepts to biology.

I shall argue that Mayr did indeed abuse the terminology of ultimate and proximate causes. Providing evidence for this thesis is, however, complicated by Mayr's obscure and sometimes chaotic reasoning. This makes it rather difficult to work out how a coherent account of ultimate and proximate causes would have to look. I believe that my accomplishment in Chap. 4 is to have demonstrated how incoherent Mayr's account of ultimate and proximate causes is, and to have shown how reasonable adjustments to remove inconsistencies reveal the triviality and ultimately the arbitrariness of Mayr's account.

The fact that Mayr is comparable to the bartender selling a particular brand of beer, instead of just beer, becomes apparent as soon as it is recognised that he has no true conceptual pluralism to offer. All he does is take subsets of causal relations and rebrand them as something special without any compelling argument. In addition to this, he pushes a relativist agenda to protect his own position and arbitrarily devalue alternative views. Once this is clearly brought to light, there is in my opinion nothing left to justify Mayr's account of ultimate and proximate causes.

Another distinction between two causal concepts which has recently been the focus of philosophical debate is that between permissive and instructive causes. This terminology originated in developmental biology and is meant to contrast the different contributions that causes can make to an effect. Brett Calcott and Pierrick Bourrat have argued about how these terms translate into philosophical theory. Here I shall side neither with Calcott nor with Bourrat but with the author initiating the debate, James Woodward, who suggested that 'permissive' and 'instructive' correspond to what has been called 'low' and 'high influence' in philosophical theories. The bewilderment arising over the terms 'permissive' and 'instructive causes' is, as I try to show in Chap. 5, a consequence of a poor semantic analysis and a superficial examination of the actual biological mechanisms to which the authors in the debate are trying to apply the terminology. If my analysis is correct, there is a good connection between philosophical and biological theory in this case. Although I am not entirely convinced by Woodward's interventionism as a general theory of causation, I have to admit that this theory gets it right when talking about permissive and instructive causes.

Next (in Chap. 6), I turn to look at suggestions from philosophers who recommend their theories to biologists. One such suggestion is to distinguish production from dependence, which is similar to Hall's above-mentioned conceptual pluralism. In this case, however, I will dismiss those ideas, since I don't see how questions raised by biologists are affected by that conceptual distinction.

Finally, I turn to the related issue of omissions (Chap. 7). Causation by disconnection is ubiquitous in biological mechanisms and one may therefore ask whether some philosophical theories deliver better results than others when viewed from the biologist's perspective. Here I shall argue that the disposition-based process theory is currently the best option, while other approaches fail to model biological mechanisms satisfactorily.

The picture emerging at the end of this book is not one of a particular theory of causation triumphing over another theory, nor is it pluralism triumphing over monism or vice versa. The outcome will be a balanced view of when to turn to conceptual pluralism. The main thesis I want to bring home in this book is that—contrary to what is sometimes assumed—many problems do not even hinge on the question of pluralism. This doesn't mean that there is no room left for pluralist positions, but the range of reasonable applications for pluralism is smaller than some may have expected.

1.5 What Not to Expect From This Book

I have used the term 'pluralism' a lot up to this point. Note, however, that my aim is not to argue about the *truth* of pluralism or monism as broad overarching ideas. At this point I am thinking of positions defending anti-essentialism and anti-reductionism (e.g. Dupré, 1993). To be candid, I feel uneasy about this big idea of pluralism because it sometimes appears to be put forward like a profession of faith. Take as an example Kellert et al. (2006) who defend a stance (rather than a position).

I think of my approach in this book as less dogmatic, in the sense that I explore pluralist ideas without ruling out monism. The reason why I am open to both options is that I see neither of them as clearly defined positions—at least not when one is talking broadly about theories of causation. One can delineate those terms more precisely when one speaks about causal concepts, as I try in Chap. 3, but I doubt that my analysis of the conceptual plurality of causal concepts transfers easily to other debates about some sort of monism-pluralism contrast.

Pluralism in its broad and rather vague sense seems to be the insight or thesis that a certain area of philosophical interest (e.g. causality) withstands the attempt to analyse it in terms of a universal and comprehensive theory. A problem is, however, that it is unclear what a universal and comprehensive theory might be. The success of a theory is often evaluated by reference to a list of desiderata, i.e. theories meeting those criteria are considered to be successes while those performing poorly are deemed failures.

I am sceptical of forcing a decision on causal monism or pluralism by first compiling such a list and then judging which theory scores best. This is an important argumentative strategy if one wants to test the merits of a particular theory, yet I don't think that it is the best way of discussing causal pluralism. Such an approach makes it far too easy to push through an arbitrary view.

An example of this kind of approach is perhaps Cartwright's (2004) discussion of the problems that bewilder anyone trying to develop a universal and comprehensive theory of causation. Her discussion focuses on some of the most dominant contemporary theories of causation. In her view, each account is a remarkable accomplishment, despite the fact that each account has its problems. She thinks that there is also a reason why those accounts cannot achieve what they aim to do: it is that 'the task they set themselves cannot be accomplished' (Cartwright, 2004, p. 813). According to Cartwright, one has to accept the possibility that one cannot have a general theory of causation, since '[t]he causal arrangements of the world may be indefinitely variable' (Cartwright, 2004, p. 818). Strictly speaking, Cartwright doesn't conclude that causal pluralism is true. She doesn't use the word 'pluralism' (in that article). Nevertheless, others regard her conclusion as a pluralist outcome (Hitchcock, 2007; Godfrey-Smith, 2009).

Another example is Hall's (2004) defence of two concepts of cause (where one doesn't find the word 'pluralism' either). Hall examines a Lewis-style counterfactual theory of token-event causation. His thesis is that a universal theory of counterfactual token-event causation cannot be formulated, since it is impossible to reconcile the commonly presupposed theoretical desiderata. There are in his view two groups of essentially incompatible desiderata: on the one hand, there are the ideas of causation being transitive, of causation being local and of causation being intrinsic; on the other, there are the ideas that omissions are causal relata and that counterfactual dependence is constitutive of causation. Counterfactual analyses of token-event causation that are meant to satisfy all of these desiderata at once are in Hall's opinion 'doomed to failure' (Hall, 2004, p. 226).

Both Cartwright's and Hall's discussions contribute to a better understanding of the limits of theories of causation. This is a reason why one should value their efforts at discussing causality. However, it is a completely different matter to *expect* them to solve the question of whether pluralism is true or not. Even if they wanted to, it is perhaps a question too big for anyone to answer.

Although there are certain desiderata that are typically mentioned in the debate on causation, one would be hard-pressed to deliver *a comprehensive list of universally (or nearly universally) accepted desiderata*. Just consider the following examples: there is on the one hand the belief in transitivity (Lewis, 1986), and on the other the belief in non-transitivity (McDermott, 1995); on the one hand the belief in contiguity (Hume, Treatise), and on the other the belief in non-contiguity (Woodward, 2003); on the one hand the belief that the right thing to do is to reduce causality (Lewis, 1986), and on the other the belief that reduction is non-mandatory (Woodward, 2003). These are only a few examples of controversy.

It would appear that in order to settle a dispute about pluralism, these controversies would have to be settled first. But could one succeed? I think there is a good chance of ending up back at the starting point, i.e. in disagreement. A little exercise in scepticism is perhaps the best way to make this more clearly visible.

From a systematic point of view, one should distinguish three variants of the thesis that there isn't a single theory satisfying all desiderata: the first variant merely states that currently no one has proposed a single theory satisfying all desiderata; the second variant states that currently no one has proposed a single theory satisfying all desiderata, but that there may be such a theory of which no one yet knows; and the third variant states that currently no one has proposed a single theory satisfying all desiderata, and that there cannot be any such theory.

Obviously, there are certain strategic decisions that would allow one to defend or refute *all* of the above theses. A simple way of refuting each of those theses is to decide that there are only a few essential desiderata. A clever decision would be to select only those desiderata that are already satisfied by some theory. In other words, a universal and comprehensive theory of causation is easy to come by if one is prepared to give up enough theoretical goals.

Other strategic decisions would allow any one of the three theses to be defended. Adopting a number of desiderata that no known theory satisfies would easily defend the first thesis. The second thesis is similarly defended. The decision must be to increase the number of desiderata until no known theory satisfies all of them. However, one must also be careful not to select desiderata that are in open conflict with each other. For instance, if there were two desiderata that are for logical reasons incompatible it wouldn't be an open question to ask whether it is possible to find a theory that satisfies all desiderata. It is then also clear how to defend the third thesis. One has to adopt at least two desiderata of which it is plainly impossible that any *single* theory can satisfy them.

The sceptical conclusion is that any view is tenable if one conceives of causal pluralism as the thesis that causal theories fail or satisfy a list of desiderata. Of course, a natural reply to this argument would be that the sceptic is mistaken in suggesting that one can choose desiderata as freely as presupposed in this argument: surely there are some desiderata that everyone has to accept? Maybe the answer to this question is 'yes'. However, be aware that agreement on a single desideratum doesn't decide the debate. What is needed is agreement on a whole *catalogue* rather than on a single desideratum. The only exception is perhaps a case where one has agreed on two incompatible desiderata. That would prove the impossibility of a single theory. But this, too, would be an unlikely course for the debate to take, since it would encourage questioning the validity of one or the other of the conflicting desiderata.

My view is that, although it is—strictly speaking—not impossible to proceed by systematically reviewing the validity of the various desiderata that have been proposed in the debate, it is inadvisable to do so. The problem is not so much that it would be tedious work, but that it would result in a highly unstable, i.e. speculative, conclusion. In order to arrive at the verdict one would first answer the question 'is causation transitive, yes or no?', then one would answer the question 'is causation local, yes or no?', then one would answer another question, and another one, until one would eventually add up the results. But if it is true that in this process a *single* decision to accept or refute a desideratum can turn the tide, then the whole strategy is extremely prone to producing an erroneous conclusion. Just one little error in judgement can result in complete failure. With this danger in mind, I prefer to maintain my sceptical attitude. If one gives weight to certain desiderata, causal pluralism as the failure of causal theories will turn out to be true. If one discards certain desiderata, causal pluralism in this sense will turn out to be false. This doesn't mean that there is no answer to this question. There does seem to be some remote possibility of deciding the question. In this book, however, I am not examining that question. I am offering *a critique of the misuse of causal concepts* instead.

References

Amundson, R. (2005). *The changing role of the embryo in evolutionary thought. Roots of evo-devo*. New York: Cambridge University Press.

Anscombe, G. E. M. (1971/1993). Causality and determination. In E. Sosa & M. Tooley (Eds.), *Causation* (pp. 88–104). New York: Oxford University Press (Reprinted from *Causality and determination. An inaugural lecture*, 1971, Cambridge University Press)

Battistella, M. (2002). Fomepizole as an antidote for ethylene glycol poisoning. *Annals of Pharmacotherapy*, *36*(6), 1085–1089. https://doi.org/10.1345/aph.1A397

Campaner, R., & Galavotti, M. C. (2007). Plurality in causality. In P. Machamer & G. Wolters (Eds.), *Thinking about causes* (pp. 178–199). Pittsburgh, PA: University of Pittsburgh Press.

Cartwright, N. (2004). Causation: One word, many things. *Philosophy of Science*, *71*(5), 805–820. https://doi.org/10.1086/426771

De Vreese, L. (2010). Disentangling causal pluralism. In R. Vanderbeeken & B. D'Hooghe (Eds.), *Worldviews, science and us. Studies of analytical metaphysics. A selection of topics from a methodological perspective* (pp. 207–223). Hackensack, NJ: World Scientific.

Dupré, J. (1993). *The disorder of things. Metaphysical foundations of the disunity of science*. Cambridge, Massachusetts and London, England: Harvard University Press.

Frege, G. (1986). Über Begriff und Gegenstand. In G. Patzig (Ed.), *Funktion, Begriff, Bedeutung. Fünf logische Studien* (6th ed.). Göttingen: Vandenhoek & Ruprecht.

Glennan, S. (2009). Productivity, relevance and natural selection. *Biology & Philosophy*, *24*, 325–339. https://doi.org/10.1007/s10539-008-9137-7

Godfrey-Smith, P. (2009). Causal pluralism. In H. Beebee, C. Hitchcock, & P. Menzies (Eds.), *The Oxford handbook of causation* (pp. 326–337). Oxford: Oxford University Press.

Hall, N. (2004). Two concepts of causation. In J. Collins, N. Hall, & L. A. Paul (Eds.), *Causation and counterfactuals* (pp. 225–276). Cambridge, Massachusetts and London, England: MIT Press.

Hitchcock, C. (2003). Of Humean bondage. *British Journal for the Philosophy of Science*, *54*(1), 1–25. https://doi.org/10.1093/bjps/54.1.1

Hitchcock, C. (2007). How to be a causal pluralist. In P. Machamer & G. Wolters (Eds.), *Thinking about causes* (pp. 200–221). Pittsburgh, PA: University of Pittsburgh Press.

Hume, D. (2000). *A treatise of human nature*. In D. F. Norton & M. J. Norton (Eds.). Oxford, New York: Oxford University Press. (Critical texts of the first editions of *A treatise of human nature*, books 1–3, and *An abstract of... a treatise of Human Nature*.)

Hüttemann, A. (2013). A disposition-based process-theory of causation. In S. Mumford & M. Tugby (Eds.), *Metaphysics and science* (pp. 101–122). Oxford: Oxford University Press.

Kellert, S. H., Longino, H. E., & Waters, C. K. (2006). Introduction. The pluralist stance. In S. H. Kellert, H. E. Longino, & C. K. Waters (Eds.), *Scientific pluralism* (Vol. XIX, pp. vii–xxix). Minneapolis, London: University of Minnesota Press.

Lewis, D. (1986). *Philosophical papers* (Vol. 2). New York, Oxford: Oxford University Press.

Mach, E. (1883/2012). In G. Wolters & G. Hon (Eds.), *Die Mechanik in ihrer Entwicklung. Historisch-kritisch dargestellt*. Berlin: Xenomoi Verlag. (Original work published 1883; reprint of the 7th edition published 1912)

Machamer, P., Darden, L., & Craver, C. F. (2000). Thinking about mechanisms. *Philosophy of Science*, *67*(1), 1–25. https://doi.org/10.1086/392759

Mackie, J. L. (1980). *The cement of the universe. A study of causation*. Oxford: Clarendon Press.

McDermott, M. (1995). Redundant causation. *British Journal for the Philosophy of Science*, *46*(4), 523–544. https://doi.org/10.1093/bjps/46.4.523

Psillos, S. (2010). Causal pluralism. In R. Vanderbeeken & B. D'Hooghe (Eds.), *Worldviews, science and us. Studies of analytical metaphysics. A selection of topics from a methodological perspective* (pp. 131–151). Hackensack, NJ: World Scientific.

Reichenbach, H. (1930). Kausalität und Wahrscheinlichkeit. *Erkenntnis*, *1*, 158–188. https://doi.org/10.1007/BF00208615

Russell, B. (1912–1913). On the notion of cause. *Proceedings of the Aristotelian Society*, *13*, 1–26.

Salmon, W. C. (1984). *Scientific explanation and the causal structure of the world*. Princeton, NJ: Princeton University Press.

Salmon, W. C. (2002/2018). A realistic account of causation. In M. Marsonet (Ed.), *The problem of realism* (ebook ed., pp. 106–134). London and New York: Routledge (original work published 2002). https://doi.org/10.4324/9781315185507

Woodward, J. (2003). *Making things happen. A theory of causal explanation*. New York: Oxford University Press.

Chapter 2
Theories of Causation

Abstract This chapter contains introductory material about several theory families in causal theory: regularity theory, counterfactual theory, probabilistic theory, interventionist theory, process theory and normality theory.

Keywords Causal fork · Causal process · Causal relevance · Central notion of cause · Complex regularity · Counterfactual · Difference maker · Disposition-based process theory of causation · Epiphenomenon · Interactive fork · Intervention · Inus condition · Normality · Possible world · Probability raising

This chapter introduces several theories of causation. The objective is to provide essential background information on those theories of causation that are relevant to my examination of conceptual causal pluralism in biology. Consequently, the following outline isn't meant to be complete, nor are the given accounts of individual theories meant to be exhaustive. The primary goal is to introduce several key concepts that will facilitate the discussion. A further goal is to shed some light on the motives behind the discussed theories—though here, too, I shall confine myself to the essentials. I shall try to give the reader *some* plausible reasons for accepting a particular theory. In addition, I would like to advise that I might use terms such as 'regularity theory' or 'process theory' generically, i.e. referring to a family of theories rather than to an individual theory. The theory families covered in this chapter are *regularity theory* (Sect. 2.1), *counterfactual theory* (Sect. 2.2), *probabilistic theory* (Sect. 2.3), *interventionist theory* (Sect. 2.4), *process theory* (Sect. 2.5) and *normality theory* (Sect. 2.6).

2.1 Regularity Theory

The central idea of any regularity theory of causation is that causal relations exhibit some sort of *regularity*. If, for instance, someone touches the surface of a balloon

K. Ehrenstein, *Causal Pluralism in the Life Sciences*,
History, Philosophy and Theory of the Life Sciences 25,
https://doi.org/10.1007/978-3-030-87942-6_2

with a needle, the balloon will burst. The cause (touching the balloon's surface with a needle) is followed by the effect (the bursting of the balloon). Provided there are no exceptional circumstances (e.g. the balloon being deflated etc.), the effect follows the cause *regularly*, i.e. whenever a balloon is touched with a needle the balloon bursts. However, why would anyone insist that regularity is so important? What are the motives for holding a regularity theory of causation?

It was David Hume who formulated the most influential arguments in favour of such a theory. Before Hume, it was a common view that a causal connection had to be a *necessary* connection (Clatterbaugh, 2009). But Hume believed that '[t]here are no ideas, which occur in metaphysics, more obscure and uncertain, than those of *power*, *force*, *energy*, or *necessary connexion*' (Enquiry, 7.3/SBN 62, italics in original). (He often uses the term 'necessary connexion' interchangeably with the former three expressions.) His aim was to trace the origin of the idea of a necessary connection. The conclusion at which he arrived was that there are no necessary connections between causes and effects.

If one follows Hume, the idea of a necessary connection has to be traced back to sense data. A central assumption is that ideas are copies of impressions, i.e. things that we have '*felt*, either by our external or internal senses' (Enquiry, 7.4/SBN 62, italics in original). He thinks this to be true of *all* ideas, even though it may not be obvious. Hume distinguishes 'simple ideas' from 'complex ideas' (Enquiry, 7.4/SBN 62). Simple ideas are directly related to sense data, while complex ideas are only indirectly related to sense data. Complex ideas can be 'known by definition' (Enquiry, 7.4/SBN 62). It is supposed that any definition of a complex idea leads back to simple ideas that somehow 'compose' the complex idea (Enquiry, 7.4/SBN 62). This means that, according to Hume, any idea (no matter how complex) is ultimately derived from impressions on internal or external senses.

If one accepts this presupposition that the ultimate sources are impressions from either internal or external senses, then there are two possibilities for how the idea of causality could have entered the human mind. Firstly, there is the question of whether the idea of a necessary connection could have entered through external senses. According to Hume, that is not the case. Even if one pays very close attention to (single) cause-effect relations, there is nothing perceptible that gives one the idea of a necessary connection. Hume claims that '[w]e only find, that the one does actually, in fact, follow the other' (Enquiry, 7.6/SBN 63). For instance, there is nothing to be *observed* and there is nothing to be *felt* if one billiard ball rolls into another, except that we see one ball approaching another and the other ball rolling away once they have collided. Hume's conclusion is that 'there is not, in any single, particular instance of cause and effect, any thing which can suggest the idea of power or necessary connexion' (Enquiry, 7.6/SBN 63).

Secondly, there is the question of whether one can acquire the idea of a necessary connection through internal senses. Hume considers two options: that one gets the idea of a power from reflecting upon the phenomenon that we can *control* our body by will (Enquiry, 7.9/SBN 63); or that one can get the idea of a power if one forms a new idea *within* the mind only (Enquiry, 7.16/SBN 67). Yet in both cases Hume arrives at the conclusion (which I shall mention but not discuss) that one cannot acquire the idea of a power, force, energy or necessary connection through internal senses.

The result of Hume's analysis is that there is nothing to perceive except that cause and effect 'are *contiguous* in time and place, and that the object we call cause *precedes* the other we call effect' (Treatise, 1.3.14.1/SB 155, italics in original).

At this point Hume's analysis reaches a dead end. It cannot explain how one arrives at the idea of a necessary connection. Hume's way out is to propose a psychological theory. He suggests that the idea of a necessary connection derives from the fact that human beings get used to regularities: '[t]he appearance of a cause always conveys the mind, by a customary transition, to the idea of the effect' (Enquiry, 7.29/SBN 76 f.). There are recurrent patterns in the course of nature that humans recognise and to which they react in a way *as if* causes would necessitate their effects. This psychological aspect is taken up in Hume's definition of causation as '*an object, followed by another, and where all the objects, similar to the first, are followed by objects similar to the second*' (Enquiry, 7.29/SBN 76, italics in original). While the world is thought to be entirely 'loose and separate' (Enquiry, 7.26/SBN 74), our understanding of it or our mental representation of it is not.

The view developed here is a strictly empirical one. According to Hume, it is impossible to tell whether two objects stand in a cause-effect relation 'without consulting experience' (Treatise, 1.3.15/SB 173). A single exposure to a cause-effect relation is not enough. Returning to our initial example, this means the following: I may witness how a balloon is touched with a needle and how the balloon bursts. But unless I have observed many instances of a needle approaching a balloon and the balloon bursting I am not entitled to draw the conclusion that the approaching needle is a cause and the bursting balloon its effect. In Hume's opinion, cause-effect relations are never revealed in a single instance.

Since he thought it impossible to recognise causes and effects in a single encounter, Hume laid down several rules that help to identify cause-effect relations. Here I am quoting only the first three of those 'general rules' (Treatise, 1.3.15/SB 173):

1. The cause and effect must be contiguous in space and time.
2. The cause must be prior to the effect.
3. There must be a constant union betwixt the cause and effect. 'Tis chiefly this quality, that constitutes the relation. (Treatise, 1.3.15/SB 173)

The complete list encompasses eight items. But it is usually only the first three that are considered to be the essential outline of Hume's position.[1] One may then summarise the Humean regularity theory of causation as the view that there is contiguity between cause and effect in space and time, that the cause has temporal priority to the effect, and that effects follow their causes regularly without exception.

Traditional interpretations of Hume's position also emphasise the thesis that *there are no necessary connections* in nature (e.g. Beebee, 2011). This view is called 'Humean metaphysics'. However, regularity theory is only one way of upholding

[1] Compare, for instance, Psillos (2002, p. 19).

such a view. Counterfactual theory, which I will outline in the subsequent Sect. 2.2, is another approach that tries to uphold the same metaphysical view. Humean metaphysics is in this respect systematically independent of regularity theory.

A question that has been debated in recent decades is whether Hume actually held the position usually ascribed to him. Some scholars think that Hume was a sceptical realist and not an anti-realist. According to their interpretation, Hume believed in necessary connections after all. It is suggested that he doubted only the possibility of knowing anything about those connections. This view is known as the 'New Hume interpretation' (Richman, 2007, p. 1).

By contrast, the outline I have provided above follows what one may call the 'orthodox' view. Although the New Hume interpretation offers interesting perspectives, it is the traditional interpretation that renders most of the debate on regularity theory intelligible. In what follows I shall, therefore, assume that Hume held the view that causal relations are nothing more than regularities.

If one follows the traditional interpretation of Hume's work, it soon turns out that the theory as Hume outlined it is too simple. One problem concerns the regularity between cause and effect. John Stuart Mill pointed this out:

> It is seldom, if ever, between a consequent and a single antecedent, that this invariable sequence subsists. It is usually between a consequent and the sum of several antecedents; the concurrence of all of them being requisite to produce, that is, to be certain of being followed by, the consequent. In such cases it is very common to single out one only of the antecedents under the denomination of Cause, calling the others merely Conditions. [...] The real Cause, is the whole of these antecedents; and we have, philosophically speaking, no right to give the name of cause to one of them, exclusively of the others. (System of Logic, Bk. III, Chap. V, § 3)

The problem raised here is as follows: What is called 'the cause' is often insufficient to bring about the effect. I may touch the surface of a balloon with a needle without the balloon bursting apart. If the balloon is deflated, the balloon will not burst. A certain pressure inside the balloon is a prerequisite for the balloon's bursting. The touching of the balloon with a needle alone is not enough.

Mill gives the example of someone who dies from a dish that he has eaten. Although it may be true in the ordinary way of speaking that the dish caused the person's dying, it would not be true to claim that the dish alone is sufficient to bring about death. Other conditions such as 'a particular bodily constitution' or 'a particular state of present health' need to be in place, too (System of Logic, Bk. III, Chap. V, § 3). Only an exhaustive list of such factors—possibly even including 'a certain state of the atmosphere' (System of Logic, Bk. III, Chap. V, § 3)—would give one the cause in the strict theoretical sense, i.e. the antecedence on which the consequent follows without exception.

A *further* problem also pointed out by Mill is that it is implausible to suggest that cause-effect relations hold exclusively between certain cause and effect types. To employ the above example again, imagine that I approach the balloon with the flame of a candle instead. This, too, will cause the balloon to burst. The bursting of the balloon (effect) is not exclusively related to the touching with a needle (cause).

The bursting of the balloon (effect) is also related to the approach of a candle flame (cause).

Hume, by contrast, thought that 'the same effect never arises but from the same cause' (Treatise, 1.3.15.6/SB 173). Mill refuted this thesis. He pointed out that 'it is not true that the same phenomenon is always produced by the same cause' (System of Logic, Bk. III, Chap. X, § 1). '[M]any causes may produce death' (System of Logic, Bk. III, Chap. X, § 1): stabbing, poisoning, suffocating or burning, for instance. Mill referred to this with the term 'Plurality of Causes' (System of Logic, Bk. III, Chap. X, § 1).

It is important to distinguish this latter problem from the former. The first issue raises the question of whether the supposed cause is sufficient for the effect to occur. Here Mill's concern is that it is not enough to isolate a single factor as 'cause'. According to Mill, a cause in the strict, i.e. in the philosophical, sense can only be a 'sum of several antecedences' (see above), i.e. a conjunction of factors. Note that this decision to switch from a single factor to a conjunction of factors does *not* give up the idea of regularity! Mill still supposes that causation is a sort of regularity. But, according to Mill, this regularity holds between a *conjunction of factors* and an effect.

The second issue concerns the question of whether such a conjunction of factors is also necessary for an effect to occur. Here Mill's concern is to point out that a conjunction of factors seldom (if ever) is necessary for the effect to occur. Other conjunctions, too, bring about the same effect. There are, for instance, at least two conjunctions of factors that bring about the bursting of a balloon. There is a conjunction where one *part* is (inter alia) the factor that I touch the balloon's surface with a needle and there is the conjunction where one *part* is (inter alia) the factor that I approach the balloon's surface with a flame. This plurality of conjunctions is what is at the heart of Mill's plurality of causes thesis. It is a thesis about the plurality of conjunctions to which the occurrence of an effect is linked. It is *not* a thesis about the 'plurality of factors' that one needs to stitch together in order to arrive at an antecedence that is followed by the effect. Or, to put it differently, Mill's plurality of causes thesis is *not* about the arbitrary selection of a factor among other factors. One may arbitrarily decide to select the eating of the dish as the cause of a person's death where one could also have selected the bodily constitution of the person who dies. But this is not what Mill has in mind when he speaks about the 'plurality of causes'. It is rather that one can rephrase Mill's plurality of causes thesis as the claim that there are many alternative conjunctions of antecedences that bring about the same (type) effect. The (type) effect death, for instance, is linked to the conjunction of factors 'eating a certain dish plus having a certain bodily constitution plus further factors' and the alternative conjunction 'getting stabbed plus having a certain bodily constitution plus further factors'.

In order to distinguish the two issues I shall speak of the 'Millian arbitrary selection principle' and the 'Millian plurality of causes principle'. The 'Millian arbitrary selection principle' is about choosing between the eating of a certain dish and the having of a certain bodily constitution. The 'Millian plurality of causes principle' is about the fact that death is brought about in many ways. In making that distinction I am following a reading of Mill that can be found in both Hart and Honoré (1985, pp. 19 f.) and Mackie (1980, pp. 60 f.). Hart and Honoré distinguish between the 'complexity' and the 'plurality' of causes. The former refers to what I have called 'arbitrary selection' and the latter refers to what I have also called 'plurality of causes'. Mackie codifies the distinction in his account of a cause as an inus condition.

'Inus condition' is an acronym that stands for 'an *insufficient* but *non-redundant* part of an *unnecessary* but *sufficient* condition' (Mackie, 1980, p. 62, italics in original). The condition is used in Mackie's account of a complex causal regularity. The account of a complex causal regularity both incorporates Mill's view of causation and adds further improvements. The inus condition gets developed via seven intermediate steps.

Firstly, Mackie suggests that a regularity holds, as Mill suggests, between a conjunction of factors (A, B and C) and the effect (P): 'All ABC are followed by P' (Mackie, 1980, p. 61). As it stands, ABC would be a *sufficient* condition for P. Secondly, Mackie accommodates the Millian plurality of causes principle. P might be caused by many different causes. Death, for instance, might be caused by stabbing or by poisoning, given certain further factors. If ABC is the conjunction that includes stabbing among the factors under which death occurs, and DGH is the conjunction that includes poisoning among other factors under which death occurs, then both conjunctions are part of a disjunctive antecedence such that 'All (ABC or DGH) are followed by P' (Mackie, 1980, p. 61). Neither ABC nor DGH is necessary for P. Either stabbing or poisoning a person would be enough to bring about death. Death does not require both stabbing and poisoning. Hence, both ABC and DGH are *unnecessary* but *sufficient* conditions. Thirdly, every part of the conjunction ABC or DGH is *non-redundant*. This means that each conjunction (as a whole) is minimally sufficient. If any part of a conjunction were omitted, the conjunction would no longer be sufficient. A missing part could be a certain physical constitution: if a person is in very good health, she might not die after poisoning (depending on the kind of poison). This means, fourthly, that any part of the conjunctions ABC or DGH on its own is *insufficient*. And this means *eventually* that A, B, C, D, G and H on their own are inus conditions of P.

But this is not yet a full account of causal regularity. It continues, fifthly, with the introduction of negative conditions that are written as capital letters with macrons in Mackie's account. '$\bar{C}$', for instance, could stand for the absence of an antidote if someone gets poisoned (Mackie, 1980, pp. 62 f.). In the sixth step, Mackie makes his account time symmetric. This means that if 'all ($AB\bar{C}$ or $DG\bar{H}$ or $JK\bar{L}$) are followed by P' then 'all P are preceded by ($AB\bar{C}$ or $DG\bar{H}$ or $JK\bar{L}$)' (Mackie, 1980, p. 63). And finally, Mackie introduces a causal field (F). The causal field is something that constitutes 'a background against which the causing goes on' (Mackie, 1980, p. 63). An example of such a causal field would be a 'block of flats as normally used and lived in' (Mackie, 1980, p. 35, italics removed). The idea of a causal field is that it states normal conditions. Anything abnormal, e.g. 'a gas leak' (Mackie, 1980, p. 34), is not included in the causal field. A causal field is—it must be emphasised—not a part of the cause. Causal regularity, according to Mackie, is then:

> In F, all ($AB\bar{C}$ or $DG\bar{H}$ or $JK\bar{L}$) are followed by P, and, in F, all P are preceded by ($AB\bar{C}$ or $DG\bar{H}$ or $JK\bar{L}$). (Mackie, 1980, p. 63)[2]

Mackie suggests that 'what is typically called a cause' in daily life 'is an inus condition or an individual instance of an inus condition' (Mackie, 1980, p. 64). The fact that the inus condition gives one some freedom to arbitrarily select a salient cause incorporates what I have called the 'Millian arbitrary selection principle'. Hence, the complex causal regularity obeys both the Millian plurality of causes principle and the Millian arbitrary selection principle.

Regularity theory is considered to be a paradigmatically type-level theory of causation, which means that its focus is on relations between types instead of tokens. As seen above, one of Hume's central theses is that it is impossible to identify a cause-effect relation in a single instance. A token cause followed by a token effect can in itself never reveal that it caused the effect. Only the detour to the type level of causation can give one that information. The proponent of regularity theory maintains that a singular (token) causal claim must always entail the truth of a covering general (type) causal claim.

2.2 Counterfactual Theory

The central feature of a counterfactual theory of causation is the analysis of causation in terms of counterfactual dependence. The starting point of such an analysis is to point out that it is intuitively persuasive to claim that effects depend counterfactually on their causes: for instance, if I had not touched the surface of the balloon with a needle the balloon would not have burst apart.

An important motive for pursuing a counterfactual analysis or a conditional analysis (as it is called by Mackie) is to employ counterfactual dependence as a means of distinguishing causal from non-causal sequences (Mackie, 1980, p. 29). That there is such a problem had been recognised by Thomas Reid who pointed out that there are paradigmatically regular but non-causal successions such as that between night and day (Reid, 1843, p. 258). There is no exception to the regular succession of night and day but one would not want to say that night is a cause of day nor that day is a cause of night.

Mill's response to this problem was to claim that '[t]he cause is not the invariable antecedent, but the *unconditional* invariable antecedent' (System of Logic, Bk. III, Chap. V, § 6, italics in original). Mill argued that '[w]e do not believe that night will be followed by day under all imaginable circumstances, but only that it will be so

[2] The second opening bracket in the quote does not occur in the original. Since its omission appears to be a misprint, the bracket is added to the quote.

provided the sun rises above the horizon' (System of Logic, Bk. III, Chap. V, § 6, italics in original). If the sun did not rise above the horizon, there would be eternal night; and if the sun did not set, there would be eternal day. Hence, Mill reasoned, the succession between night and day is conditional and, therefore, non-causal.

Mackie, however, criticised Mill for not distinguishing '*de facto* unconditional regularity from counterfactually unconditional regularity' (Mackie, 1980, p. 81). The succession of night and day is *de facto un*conditional. Given the actual circumstances, there are no exceptions to the succession of night and day. In order to distinguish causal from non-causal sequences another sort of conditional is required: one would have to imagine the *contrary-to-fact* circumstance that the earth stands still. For Mill's response to work, one would have to entertain contrary-to-fact conditionals, i.e. counterfactual conditionals. But here Mackie saw the problem that 'the holding of a counterfactual conditional is not a fully objective matter' (Mackie, 1980, p. 82). In order to assess the truth of a counterfactual conditional one has to *imagine* contrary-to-fact circumstances, which makes it uncertain that counterfactual conditionals can have objective truth values. In accordance with this, Mackie confined his conditional analysis to the *meaning* of causal statements. Mackie conceded that causal statements do often suppose some counterfactual truth but he denied that one could objectively distinguish causal from non-causal sequences by means of a counterfactual analysis (Mackie, 1980, Chap. 2).

A breakthrough in the development of a counterfactual theory of causation was the development of possible-worlds semantics and its employment as a means of analysing causation. David Lewis (1973; reprinted in Lewis, 1986) developed the most influential account of this sort. His starting point is what he calls 'Hume's "other words"' (Lewis, 1986, p. 160). Although Hume didn't offer a counterfactual theory of causation, Hume did offer an alternative definition of cause that is phrased in terms of counterfactual dependence:

> [W]e may define a cause to be *an object, followed by another, and where all the objects, similar to the first, are followed by objects similar to the second*. Or in other words, *where, if the first object had not been, the second never had existed*. (Enquiry, 7.29/SBN 76, italics in original)

While the first definition adheres to the regularity view of causation, the second definition points to an understanding of causation in terms of counterfactual dependence. One of the chief problems that Lewis had to overcome in order to make this second approach work was to show that 'counterfactuals need not remain ill understood' (Lewis, 1986, p. 161). Lewis suggested that one can analyse causal dependence in terms of counterfactual dependence. Causal dependence holds between events. Counterfactual dependence holds between propositions. Lewis stipulated that one can pair any event with a corresponding proposition so that one may use the same means for the analysis of causal dependence that one uses for the analysis of counterfactual dependence (Lewis, 1986, p. 165 f.).

Suppose that there are the propositions A and C and the operator $\Box\!\!\rightarrow$. If put together, one has the counterfactual $A \Box\!\!\rightarrow C$, which reads as 'if A were true, then C would also be true' (Lewis, 1986, p. 164). For instance, if it were true that I touched the surface of a balloon with a needle, then it would also be true that the balloon would burst. The event that the balloon bursts depends on the event that I touch the balloon with a needle. Between the events there is causal dependence and between the propositions that describe the occurrence of these events there is counterfactual dependence. At least, this is what appears to be intuitively true. Giving a theoretical account of the truth conditions of the counterfactual is more demanding. In order to put Lewis's account into operation, one first needs to introduce the notion of a possible world, and then the notion of comparative similarity.

One of Lewis's suppositions is that there is the actual world and there are *also* other so-called possible worlds. The actual world is simply our world. Possible worlds are supposed to be 'more things of *that* sort, differing not in kind but only in what goes on at them' (Lewis, 2001, p. 85, italics in original). In this world, for instance, Angela Merkel became chancellor of Germany. But, in another world with a different history, she may never have held public office at all—or she may have served an even greater number of terms as chancellor. Lewis defends a realism about these worlds, i.e. he supposes that those worlds in fact *exist*. According to that view, the 'actual world is only one world among others' (Lewis, 2001, p. 85). The only reason why it is called 'actual' is that we happen to inhabit it. In this sense the term 'actual' merely points to our position among the possible worlds. It doesn't mean that our world is in any sense more real than the other worlds (Lewis, 2001, pp. 85 f.).

A further ingredient of Lewis's account is the notion of 'comparative similarity' (Lewis, 1986, pp. 163 f.). The evaluation of the truth value of counterfactuals requires worlds to be compared against each other. But how is that done? According to Lewis, there is not so much to explain. He supposes that similarity among possible worlds is a primitive and an overall matter: it is thought to be *primitive* in the sense that we already have an intuitive understanding of comparative similarity and it is thought that (according to this intuitive understanding) similarity is treated as an *overall matter*. Lewis maintains that '[w]e do make judgements of comparative overall similarity—of people, for instance—by balancing off many respects of similarity and difference' (Lewis, 1986, p. 163).

Although Lewis supposes that we are already familiar with comparative overall similarity, it is instructive to contrast this notion of similarity with what it is *not*. One may judge, for instance, that Peter is more similar to Petra than to Paul *or* we may judge that Peter is more similar to Paul than to Petra. Which judgements one prefers depend on the way in which one weighs different factors against each other. Perhaps Peter and Petra share a liking for a certain sort of ice cream but Paul doesn't have the same liking. Then one would consider Peter and Petra more similar to each other than to Paul *with respect* to that liking of ice cream. But if one believes another factor to be more important the judgement could be different. If Peter and Paul share a common hair colour and Petra does not, then Peter and Paul could be considered to be more similar to each other than to Petra. Examples such as these suggest that similarity judgements are often *relativised* to certain factors that are considered to be important.

However, this is *not* what Lewis has in mind. The concept of comparative *overall* similarity is supposed to take *all* factors into account and weigh them against each other. In other words, overall similarity is *not relativised* to specific characteristics such as an ice-cream preference. It is meant to be a *balance* between all the characteristics the individuals possess. For Peter, Paul and Petra it would have to be a balance between all their characteristics, such as hair colour, nationality, age, sex, favourite movies, height, bravery, number of siblings, and so on. When faced with a question that demands a judgement of comparative overall similarity, one is supposed to give a *non-relativised* answer. One may not reply, for instance, 'Peter is more similar to Petra than to Paul with respect to height.' Instead one is supposed to give an answer such as the following: 'Peter is more similar to Petra than to Paul (full stop).'

Lewis suggests that one need not specify 'how to balance the respects of comparison against each other' (Lewis, 1986, p. 163). In his opinion 'our mutual expectations about the weighting factors are definite and accurate enough to permit communication' (Lewis, 1986, p. 163). However, Lewis formulates some of *his* expectations explicitly when it comes to the comparison of worlds. According to Lewis, two factors are important: there are, firstly, differences with respect to the spatiotemporal distribution of fact and, secondly, differences with respect to law.[3] A difference with respect to the spatiotemporal distribution of fact would be that Cologne Cathedral has two spires in this world but only one spire in another world. A difference with respect to law would be that a law in this world holds without exception but there is a violation of that law in another world. In this world, for instance, my teacup falls to the ground if pushed over the edge of the table but in another world the teacup may hover in mid-air, even though everything else in that world falls to the ground. According to Lewis the two kinds of difference trade off against each other. Lewis is often willing to sacrifice the integrity of law in favour of similarities in the distribution of facts. 'Comprehensive and exact similarities of particular fact throughout large spatiotemporal regions' have by default more weight than a minor violation of law, a so called 'miracle' (Lewis, 1986, p. 164).

Having explained the notion of a possible world and the notion of comparative similarity, I am now in a position to spell out the truth conditions of the counterfactual $A \Box\!\!\rightarrow C$. According to Lewis, the conditions are as follows:

> $A \Box\!\!\rightarrow C$ is true (at a world w) iff either (1) there are no possible A-worlds (in which case $A \Box\!\!\rightarrow C$ is *vacuous*), or (2) some A-world where C holds is closer (to w) than is any A-world where C does not hold. In other words, a counterfactual is nonvacuously true iff it takes less of a departure from actuality to make the consequent true along with the antecedent than it does to make the antecedent true without the consequent. (Lewis, 1986, p. 164, italics in original)

[3] Lewis also provides a more detailed catalogue (Lewis, 1986, pp. 47 f.). However, the more details are made explicit, the more technical aspects enter into the account, undermining the idea of primitiveness.

Take the original example once more: If it were true that I touched the surface of a balloon with the tip of a needle, then it would also be true that the balloon would burst. The first step in the evaluation of the counterfactual is to check whether it is vacuously true. This, however, is not the case. There are worlds where A is true, i.e. there are worlds in which I touch the surface of a balloon with the tip of a needle. Hence, one cannot say that the counterfactual is vacuously true. Since the question is not yet settled, I proceed to condition two and check whether it is fulfilled. An A-world is a world in which I touch the surface of a balloon with the tip of a needle. In some of these A-worlds it is also the case that C, i.e. it is the case that the balloon bursts. In other A-worlds it is not the case that C, i.e. it is not the case that the balloon bursts. The question that one needs to answer is whether the former or the latter A-worlds are more similar to our world, i.e. the actual world.

Worlds in which both A and C are true are very similar to our world: they have similar laws and a similar distribution of fact. Worlds where A but not C is true are less similar: there are violations of law (e.g. the balloon not bursting despite being punctured) or there are dissimilarities in fact (e.g. the balloon not being inflated). Worlds where both A and C are true are more similar to the actual world than worlds where A but not C is true: there is no A-and-not-C world that is more similar to actuality than an A-and-C world. Therefore, I conclude that the counterfactual is true at our world.

Causal dependence requires that two counterfactuals are true:

$O(c) \,\Box\!\!\rightarrow O(e)$ and $\sim O(c) \,\Box\!\!\rightarrow\, \sim O(e)$ (Lewis, 1986, p. 167)

'O(c)' is the proposition that the cause occurs and 'O(e)' is the proposition that the effect occurs. I have just shown that in the example the former counterfactual is true. It should be obvious that the latter is also true: if it were true that I do not touch the balloon, it would be true that the balloon does not burst. Hence, there is causal dependence between the events.

A more interesting problem, of course, is that some regularities, such as the succession of night and day, are non-causal. Cases such as these are not straightforwardly resolved, even on a counterfactual account. Imagine that there is an event c that causes first e and along a different causal path *and* a short time later f (see Fig. 2.1). If temporal succession were the only criterion, it would look as if c had caused e and e in turn had caused f, although f is caused directly by c. In this causal structure e is an epiphenomenon. It doesn't cause f, but it appears as if it causes f.

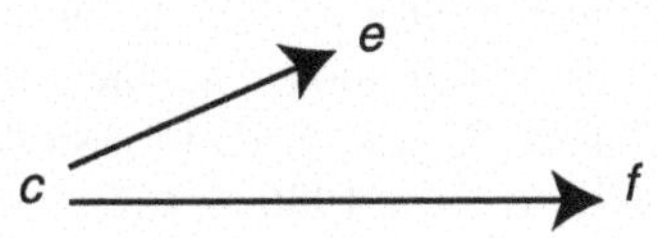

Fig. 2.1 Illustration of a causal structure in which an epiphenomenon (e) occurs. This is my illustration. It is not shown in Lewis (1986)

Regularity theory has a problem. However, counterfactual theory is also confronted with a problem: If one supposes that there is causal dependence between c and e, then one may conclude from the fact that e has occurred that c must have occurred, too. But if c, too, must have occurred, one may draw the further conclusion that f, too, must have occurred. Eventually, the conclusion is that '[w]e have a spurious causal dependence of f on e' (Lewis, 1986, p. 170).

At this point Lewis's suggestion is 'to deny the counterfactuals that cause the trouble' (Lewis, 1986, p. 170). Lewis argues that a world where e is missing but c is still there causing f is more similar to actuality than a world where c is also missing. Obviously, one has to assume that at some point one or several laws are broken (if one assumes determinism, as Lewis does). This, however, is in Lewis's opinion outweighed by the prolonged 'spatiotemporal region of perfect match between our actual world and the selected alternative' (Lewis, 1986, p. 171). If one follows this reasoning, f depends causally on c without also depending on e. These forbidden counterfactuals are also known as 'backtracking, or back-and-then-forward, counterfactuals' (Lewis, 2004, p. 78).

Lewis's strategy would supply the following solution to the initial problem: if c were the event that the earth rotates, e the event that there is night, and f the event that there is day, then night and day would causally depend on the rotation of the earth but day would not causally depend on night, nor would night on day.

One should note that counterfactual theory, although it offers a solution to the problem of epiphenomena, faces an analogous problem with counterfactuals. As pointed out by Jaegwon Kim, there are various examples of *non*-causal counterfactual truths, e.g. 'If I had not written "r" twice in succession, I would not have written "Larry"' (Kim, 1973, p. 571). In order to deal with this and other counterexamples of that sort, Lewis provides a theory of events as a supplement to his theory of causation (Lewis, 1986, pp. 241–269). The solution to the above problem is to demand that causal dependence requires counterfactual dependence of *distinct* events (Lewis, 1986, p. 259). This requirement is not fulfilled in the case of writing 'Larry'. The event *writing 'r' twice in succession* is not distinct from the event *writing 'Larry'*, because the former is a part of the latter. Hence, the two events overlap and are not distinct.

A weak point of the counterfactual analysis is preemption. One has a preemption case when there is an effect e that was caused by c_1 but if c_1 hadn't been there another cause c_2 would have caused e. The problem with such a case is that the counterfactual $\sim O(c_1) \Box\!\!\rightarrow \sim O(e)$ isn't true. Hence, causal dependence is broken. Lewis's response to this problem is to treat causation as an ancestor of causal dependence: causal dependence implies causation, but not vice versa. One cannot have causal dependence without causation, but one can have causation without causal dependence. Furthermore, Lewis needs to assume that causation (but not causal dependence) is transitive (Lewis, 1986, p. 167).

These features of the theory allow Lewis to deal with cases of early preemption. His strategy is to postulate that neither c_1 nor c_2 causes e directly. It is rather that there are two causal chains with intermediates both leading to e. If the one causal chain beginning with c_1 interferes with the other causal chain beginning with c_2

and aborts it before it can culminate in the effect *e*, only the causal chain beginning with c_1 gets completed. This allows c_1 to be identified as the cause of *e* (Lewis, 1986, pp. 171 f.). Other preemption cases have proven to be more difficult to resolve. Serious problems arise, for instance, if one faces scenarios where there are no causal chains to be interfered with. These problems with redundant causes led Lewis to a revision of his original theory (Lewis, 2004). However, at this point I shall not discuss late preemption, although I will return to it in Sect. 5.2.

Here it is more important to point out certain features of the theory that are significant from a systematic perspective. The first is that Lewis's theory articulates token-level causal claims. He states clearly that he analyses 'causation in particular cases' and not 'causal generalizations' (Lewis, 1986, pp. 161 f.). The second is that Lewis is also explicit about his intention to analyse 'a broad and nondiscriminatory concept of causation' (Lewis, 1986, p. 162). This means that he is interested in the analyses of the concept of '*a* cause' rather than '*the* cause'. Hitchcock calls this the 'egalitarian picture of causation' (Hitchcock, 2007, p. 203). This contrast will be clearer once I have introduced normality theory in Sect. 2.6. For the time being, I shall just mention that Lewis explicitly states what is only implicit in other approaches. It is in fact common among philosophers to analyse the concept of '*a* cause' rather than '*the* cause', thus it is often not stated explicitly that an approach is meant to be non-discriminatory.

Another philosophically significant feature of this analysis is that it, too, carries out the Humean programme. Lewis presupposes 'the doctrine that all there is to the world is a vast mosaic of local matters of particular fact, just one little thing and then another' (Lewis, 1986, p. ix). According to this view, all causal facts supervene on the Humean mosaic. This means that once the non-causal facts are fixed, the causal facts, too, are fixed. If it were true that causal facts supervene on non-causal facts, any difference in causal facts would imply a difference in non-causal facts, although there is no implication in the reverse direction. The same kind of causal fact may supervene on different kinds of non-causal fact. Lewis's overall philosophical programme is to defend the '*tenability*' of this view (Lewis, 1986, p. xi, italics in original).

2.3 Probabilistic Theory

According to probabilistic theories of causation, causal relations are probability relations. This thesis is in stark contrast to conceptions of causation as a necessary connection or an invariable (complex) regularity. So, why would anyone pursue such an approach?

If one follows the thoughts of Hans Reichenbach, probability theory (*not* the probabilistic theory of causation) first emerged as a theory of gambling, which didn't help to foster its reputation as a scientific method. In earlier times, according to Reichenbach, probability theory was seen as an admission of ignorance that should in principle be overcome by science. He criticised this stance, arguing that there are factual limits, not only on our measurement methods in quantum physics (i.e. the Heisenberg uncertainty principle) but also in the macroscopic world, which make it necessary to develop a proper understanding of probability theory (Reichenbach, 1930).

Reichenbach developed his probabilistic theory of causation in the context of the question of how to account for the direction of time (Reichenbach, 1999, § 3). He argued against ancient philosophers such as Parmenides and modern philosophers such as Kant: the former was criticised for claiming that time is an 'illusion', the latter for proposing a concept of time that renders time 'subjective' (Reichenbach, 1999, pp. 4 ff.; pp. 12 ff.). Reichenbach's strategy was to reduce time to causal order (Reichenbach, 1999, p. 24). However, this approach is complicated by the fact that physics treats causal phenomena as if they were reversible in time. Equations usually postulate a causal connection without defining a causal direction (Reichenbach, 1999, p. 29). The trajectory of a ball, for instance, can be described in either time direction (Reichenbach, 1999, pp. 30 f.).

In order to reduce the direction of time to causal order Reichenbach first introduces the principle of local comparability of time order. This principle stipulates that spatiotemporally close processes run in the *same* temporal direction. Once the direction of one process is known, one can use this principle as a means of deducing the direction of other processes (Reichenbach, 1999, pp. 34 ff.).

The direction of time is deduced in different ways in Reichenbach (1999). One account is to compare temporarily[4] isolated systems with surrounding systems. This allows Reichenbach to define positive time as '[t]he direction in which most thermodynamical processes in isolated systems occur' (Reichenbach, 1999, p. 127). Another account is to deduce the direction of time using the principle of the common cause.[5]

The core idea of Reichenbach's principle of the common cause is that where 'an improbable coincidence has occurred, there must exist a common cause' (Reichenbach, 1999, p. 157, italics removed). In everyday situations this would mean that the simultaneous failure of several lights in a room warrants the conjecture that the power supply was disconnected. The coincidental failure of all lights for individual reasons would be regarded as improbable (Reichenbach, 1999, p. 157).

That a coincidence is more frequent than expected is stated formally by Reichenbach in the form of the inequality

$$P(A\&B) > P(A)P(B)\,, \tag{2.1}$$

[4] This presupposes a time order but no time direction.

[5] Reichenbach notes that this doesn't introduce a new assumption. He claims that this principle is derivable from thermodynamical considerations (Reichenbach, 1999, p. 157).

which means that the probability of the joint occurrence of the events A and B is higher than the product of the probabilities of the individual occurrences of A and B (Reichenbach, 1999, p. 158).[6]

Reichenbach then defines a conjunctive fork, in which C causes both A and B but A does not cause B and vice versa. In a conjunctive fork the following conditions have to be fulfilled (Reichenbach, 1999, p. 159):

$$P(A\&B|C) = P(A|C)P(B|C)\ , \tag{2.2}$$

$$P(A\&B|{\sim}C) = P(A|{\sim}C)P(B|{\sim}C)\ , \tag{2.3}$$

$$P(A|C) > P(A|{\sim}C)\ , \tag{2.4}$$

$$P(B|C) > P(B|{\sim}C)\ . \tag{2.5}$$

The inequality (2.1) can be derived from (2.2)–(2.5). C is assumed to be greater than zero and smaller than one (Reichenbach, 1999, pp. 159 f.).

Reichenbach uses the conjunctive fork for a definition of time direction. He argues that although A and B might have a common effect E (which could also satisfy the conditions (2.2)–(2.5) if we substituted C with E), it would be impossible for E to exist without A and B also having a common cause C. The reason Reichenbach offers is that E would otherwise turn out to be a final cause. However, according to his earlier discussion of thermodynamics, the existence of final causes would here be ruled out. In Reichenbach's terminology the fork consisting of A, B and E is open to the past, while the fork consisting of C, A and B is open to the future. Since the former fork cannot exist in the absence of the latter fork (which would constitute a double fork consisting of C, A, B, and E), Reichenbach concludes that the only fork that can exist as an open fork *alone*, i.e. without being part of a double fork, must be a fork that is open to the future, thus defining the direction of time (Reichenbach, 1999, pp. 161 f.).

Already present in Reichenbach's account of a causal fork, although not yet named as such, is the idea that a third variable may screen off other variables (conditions (2.2) and (2.3)). Reichenbach introduces this terminology in his account of the relation of being causally between. In a causal chain of three successive events the intermediate event 'screens off' the last event in the sense that once it is known how the intermediate event contributes to the last event the first event becomes irrelevant for the prediction of the last event. As Reichenbach puts it, the contribution of the first event gets 'absorbed' into the intermediate event. At this point he introduces the familiar expression 'to screen off'. In this case the intermediate event screens off the first event from the last event (Reichenbach, 1999, p. 189). The same happens in common cause structures. The storm and the

[6] Here and below I modify Reichenbach's notation so that it conforms to present-day conventions. An important source of confusion is that he writes conditional probabilities in a different order: while it is now common to write '$P(B|A)$', Reichenbach wrote '$P(A, B)$'. He explained that notation in Reichenbach (1949).

low reading on the barometer are both caused by a drop in atmospheric pressure. Although one might use the barometer reading (in practice) for a prediction of a storm, the event that the atmospheric pressure fell screens off the barometer reading from the occurrence of the storm (Reichenbach, 1999, p. 193).

Reichenbach also considers that one may obtain an order and direction of time by introducing marks. He describes a mark as 'the result of an intervention by means of an irreversible process' (Reichenbach, 1999, p. 198). An example would be to put a red filter in the path of a light beam (Reichenbach, 1999, pp. 198 f.). Reichenbach suggests that the application of marks is an alternative to the principle of local comparability of time order when we want to infer time direction (Reichenbach, 1999, p. 200).

Having obtained an order and direction of time, Reichenbach eventually introduces a definition of causal relevance that presupposes the order and direction of time as given. An event A_1 that precedes event A_3 in time is causally relevant to A_3 if

$$P(A_3|A_1) > P(A_3) \tag{2.6}$$

and there are no other events $A_2^{(1)} \ldots A_2^{(n)}$ earlier or simultaneous to A_3 screening off A_1 from A_3 (Reichenbach, 1999, p. 204).

Another approach that leads into a similar account of probabilistic causality is Patrick Suppes's (1970) theory. However, an important difference is that Suppes didn't attempt to reduce the direction of time to causal order. He presupposed time direction as given, since he didn't believe that Reichenbach had successfully reduced time (Suppes, 1970, pp. 80 f.). Suppes's theory is motivated from an everyday perspective, arguing that colloquial language provides many clues to the fact that people don't think of causation as necessitation (Suppes, 1970, p. 7; p. 80).

As a starting point Suppes (1970, p. 12) introduces the notion of a prima facie cause. An event $(B_{t'})$ is a prima facie cause of another event (A_t) iff

$$t' < t\,, \tag{2.7}$$

$$P(B_{t'}) > 0\,, \tag{2.8}$$

$$P(A_t|B_{t'}) > P(A_t)\,. \tag{2.9}$$

As the name 'prima facie' indicates, there are exceptions. Some causes turn out to be spurious under this definition. Just as in Reichenbach's theory, it is necessary to demand that $B_{t'}$ isn't screened off from A_t.

Suppes provides such a solution and offers three definitions of spurious causes (Suppes, 1970, pp. 23 ff.). The idea behind these fine-tuned definitions is always to seek some earlier cause that might account 'for the conditional probability of the effect just as well' (Suppes, 1970, p. 21).

This strategy of distinguishing spurious from genuine causes with reference to time order was, however, soon challenged. Richard Otte (1981) devised various counterexamples to Suppes's theory, arguing that Suppes's definitions of spurious causes had the implausible consequence that only the earliest cause that accounts for the conditional probability of the effect would be considered to be the genuine cause.

Recapitulating one of Otte's counterexamples will help illustrate one of the problems: Imagine that there is an intact window at time t and that there are two processes running towards that window. One process is a bullet hitting the window at t', the other process is a stone hitting the window simultaneously at t'. Intuitively, both processes should be considered to be genuine causes. This is apparently a case of overdetermination. However, as soon as time order is introduced as a criterion the bullet turns out to be a spurious cause. Given that the bullet and the stone hit the window at the same time, there are circumstances possible in which the bullet is fired after the slingshot releases the stone. If both processes are initiated from the same distance to the window, the stone has to depart earlier (if it is to hit the window at the same time), since it travels more slowly. Hence, the bullet might be fired at t'', while the stone might be slung even earlier at t'''; rendering the bullet arbitrarily 'spurious' (Otte, 1981, pp. 173 f.).

Otte explored various ways of amending Suppes's theory in order to cope with such counterexamples. His final verdict was, however, that Suppes's theory is beyond repair, since Suppes-style theories are also incapable of dealing with so-called interactive forks.

Interactive forks were previously discussed by Wesley Salmon (1978), who had developed a statistical-relevance model for scientific explanation (Salmon, 1971)[7] as a replacement for Hempel's (1965) account of inductive-statistical explanation. While Hempel assumed that explanations should demonstrate how an explanandum event was to be expected with a probability close to one (Hempel, 1965, pp. 381 ff.), Salmon opted for a model that compared prior and posterior probabilities of the explanandum event relative to explanatory facts (Salmon, 1971, p. 36). In this latter model Salmon incorporated ideas from Reichenbach, specifically the notion of screening-off (Salmon, 1971, p. 55). In addition, Salmon made extensive use of Reichenbach's definition of a conjunctive fork, e.g. in his argument for scientific realism based on the various ways of determining Avogadro's number (Salmon, 1984, pp. 213 ff.). However, Salmon also delineated the limits of such reasoning more precisely through his analysis of interactive forks.

Such forks are instantiated in cases where the correlated effects A and B are *not* screened off by their common cause C. Salmon (1984, pp. 168 ff.) illustrates this with a case where we consider the interaction of two balls in a billiard game. The 8-ball is positioned where it can be played with the cue ball in such a way that it will fall into one of the far pockets. However, Salmon also stipulates that the only way to hit the 8-ball into the pocket will also ensure with a probability close to one that the cue ball will fall into a pocket as well. Hence, the idea is that there might be

[7] Note that this is not a theory of probabilistic causality (Salmon, 1971, p. 81). Consequently, interactive forks are not yet a subject of consideration in that work. Interactive forks are discussed in Salmon (1978) and a comprehensive view of probabilistic causality is laid out in Salmon (1984).

a situation in which the 8-ball cannot fall without the cue ball falling, too. Salmon assumes that hitting the 8-ball in a way that puts it into the pocket might have a probability of 0.5, i.e. there are quite a few cases where the 8-ball would be hit but not fall into the pocket. In such a scenario the conditional probability of A (the 8-ball falling into the pocket), given C (the shot is attempted) is $P(A|C) = 0.5$. The same holds for the conditional probability of B (the cue ball falling into the pocket), given C (the shot is attempted): $P(B|C) = 0.5$. In cases where C screens off A from B the Eq. (2.2) has to hold. This means the conditional probability $P(A\&B|C)$ would have to be equal to the product of $P(A|C)$ and $P(B|C)$. This is obviously not the case: the product of $P(A|C)$ and $P(B|C)$ is 0.25, while $P(A\&B|C)$ is supposed to be close to 1.[8]

A concern with this example might be that we could reanalyse the billiard game. It might then turn out that A and B are screened off by background circumstances of which we are unaware. Salmon argues, however, that interactive forks are real. He believes that there are also less disputable examples from quantum physics, such as the Compton scattering experiment (Salmon, 1978, pp. 692 ff.; Salmon, 1984, pp. 169 f.).

A further difference between Suppes's and Salmon's accounts is how they deal with cases in which it is uncertain that the cause increases the probability of the occurrence of the effect. An early counterexample of that sort was brought up by Deborah Rosen, a student of Suppes (Suppes, 1970, pp. 37). Rosen's example is to imagine a golfer who shoots a ball into a tree where it is deflected by a branch so that it falls into the cup, resulting in a so-called birdie (scoring under par). This succession of events is then further specified by estimating probabilities for the occurrence of these events. It is stipulated that the golfer's skill is low and that therefore the probability of the golfer making a birdie is low as well. The conditional probability of making the birdie, given that the ball hits the tree on its course to the hole, is estimated to be even lower. However, in hindsight it is also evident that the fact that the ball hit the tree was significant for the ball to drop into the cup (Suppes, 1970, p. 41; see also Rosen, 1978, pp. 607 f.).

Suppes calls this the problem of 'improbable consequences' (Suppes, 1970, p. 41); Salmon prefers to speak of 'negative statistical relevance', since it is not the low probability that causes the problem but the negative statistical relevance of the cause for the occurrence of effect (Salmon, 1980, pp. 62 f.; Salmon, 1984, p.193). While Suppes and others try to save the principle that all causes raise the probability of their effects, Salmon argues for the thesis that some causes have a negative statistical relevance to their effects.

There are three strategies, addressed by Salmon, that might save the principle that all causes raise the probability of their effects. One strategy, which Salmon

[8] Note that this example may also face a problem with temporal order, as discussed above: if one attempts to apply a Suppes-style account of probabilistic causality, it is conceivable that the 8-ball might fall before the cue ball, rendering it erroneously a cause of the cue ball's falling (Salmon, 1984, p. 193).

names 'the method of more detailed specification of events' and ascribes to Rosen (1978), is to maintain that we might discover that we erroneously assumed that the cause lowered the probability of the effect. If we reexamined the birdie, we might find that a different, possibly more fine-grained, description of the case leads to the conclusion that the particular circumstance raised the probability—contrary to our initial expectation (Salmon, 1980, pp. 62 ff.; Salmon, 1984, pp. 193 ff.).

Salmon discards this idea as 'thoroughly unconvincing' and likens it to attempts to restore determinism by postulating that anything will turn out to be deterministic once enough information is gathered (Salmon, 1984, pp. 194 f.).

A second strategy ascribed to I. J. Good (1961a, 1961b, 1962) and called 'the method of interpolated causal links' by Salmon (1980, p. 64) is to postulate further intermediate events of which the event immediately preceding the effect raises the probability of the effect. Applied to the golf example again, we could argue that there are at least four successive events: the shot, the travel through mid-air, the collision with the tree, and finally the drop into the cup. Without the collision with the tree the probability of the ball dropping into cup is virtually zero, since the travel through mid-air goes completely in the wrong direction. When the collision occurs, however, the probability of reaching the cup goes up, although only very slightly (Salmon, 1984, p. 195).

Salmon doesn't dispute that this would save the principle in the sense that the drop into the cup would be caused by an event that increases the probability of the effect. The difficulty he points out is that this merely shifts the problem: now the second event in that causal chain lowers the probability of the ball reaching the cup, i.e. the travel through mid-air is of negative statistical relevance instead (Salmon, 1984, p. 195).

A third strategy, considered by Salmon but also eventually dismissed by him as a universal solution, would be to try and reassess each step of the causal chain in isolation. Salmon calls this 'the method of successive reconditionalization' (Salmon, 1984, p. 196). Applied to Rosen's example, Salmon (1984, pp. 199 f.) solves the problem as follows: A is the class of shots of the golf player at that particular hole, B is the badly performed swing that sets the ball on a wrong trajectory (i.e. a 'slice' in golf terminology), C is the ball on its trajectory towards the tree, D is the collision with the tree and E is finally the drop into the hole. When we now consider the first step in this causal chain, we obtain the inequality

$$P_A(C|B) > P_A(C|{\sim}B)\ , \tag{2.10}$$

stating that the probability of the ball being on a wrong trajectory (C), given a badly performed swing (B), is higher than the probability of the ball being on a wrong trajectory (C), given the absence of a badly performed swing (${\sim}B$).

Moving on to the next step in the causal chain we shift to the class of badly performed swings (B) and get the inequality

$$P_B(D|C) > P_B(D|{\sim}C)\ , \tag{2.11}$$

meaning that the probability of a collision with the tree (D) is higher when the ball is on a wrong trajectory (C) than in the absence of a wrong trajectory ($\sim C$).

Finally, we look at the link between bouncing back off the tree and dropping into the cup. This gives yet another similar inequality for the class of off-target balls (C)

$$P_C(E|D) > P_C(E|{\sim}D) , \tag{2.12}$$

stating that the probability of the birdie (E) is higher when the ball hits the tree (D) than when it doesn't ($\sim D$). Salmon accepts that each of these statements is 'manifestly correct' (Salmon, 1984, p. 200).

Unfortunately, this strategy of viewing each link of the causal chain in isolation doesn't rescue Suppes-style theories—as Salmon is quick to point out. A problem arises with indirect causes. Suppes made it a necessary condition that an indirect cause must be a prima facie cause, too (Suppes, 1970, p. 28). This means that any indirect cause must increase the probability of the effect indirectly resulting from it. In the golf example this would mean that being on a bad trajectory (C) would have to increase the probability of the birdie (E), instead of *only* increasing the probability of a collision with the tree (D). Salmon therefore abandons Suppes's theory (Salmon, 1984, p. 200).

Salmon also thinks that his method of successive reconditionalisation is *not* always applicable; in quantum theory, for instance. Demonstrating the principle with an artificial example, he suggests considering an atom in an excited state, which may get back to lower states in two ways: it may leave the 4th energy level via the 3rd energy level, leaping the 2nd energy level on its way to the 1st energy level (still one level above the ground state, but with a probability of 1 for the further transition to the ground state); or it may leap from the 4th energy level directly to the 2nd energy level and then to the 1st energy level. A transition from 3rd energy level to 2nd energy level is assumed to be impossible. Salmon stipulates (fictitious) probabilities for these transitions in such a way that transitioning via the 2nd energy level lowers the probability of returning to the 1st energy level compared to the return to the 1st energy level via the 3rd energy level, i.e. occupation of the 2nd energy level is of negative statistical relevance (Salmon, 1980, p. 65; Salmon, 1984, pp. 200 f.).

Salmon argues that it is impossible in this scenario to restore positive statistical relevance. He maintains that there is no room for a more fine-grained description, ruling out both the method of more detailed specifications of events and the method of interpolating causal links. The method of successive reconditionalisation would also fail, since it was not possible to conditionalise accordingly. Considering atoms occupying the 4th energy level, the inequality for reaching the 1st energy level via the 2nd energy level is

$$P_4(1|2) < P_4(1|{\sim}2) . \tag{2.13}$$

Salmon suggests that one might try to utilise the non-occurrence of the occupation of the 3rd energy level

$$P_{\sim 3}(1|2) > P_{\sim 3}(1|\sim 2)\ , \tag{2.14}$$

but dismisses this for the reason that the non-occurrence of the 3rd energy level isn't a physical event preceding the occupation of the 2nd energy level (Salmon, 1980, pp. 69 f.; Salmon, 1984, p. 201).

This latter example indicates that positive statistical relevance might not be a fundamental principle of a probabilistic theory of causation. Salmon, therefore, suggests that it is the transmission of probabilistic causal influence instead (Salmon, 1984, p. 202). Following Salmon, statistical relevance relations are not constitutive of causation but have to be explained in terms of causal relations. Salmon proposes a two-tier model of scientific explanation: The first level is to identify the statistical relevance relations in which the explanandum event is embedded. The second level is to account for these statistical relevance relations in terms of causal relations (Salmon, 1984, p. 22). Salmon analyses these causal relations in terms of a process theory that I discuss in the separate Sect. 2.5.

A further problem with Reichenbach-Suppes style theories is that they can give rise to Simpson's paradox—named after E. H. Simpson (1951). While many arguments regarding Simpson's paradox rest on fictitious examples, it is known that the problem arises in real data, too (Wagner, 1982). Judea Pearl (2014)[9] traces the recognition of that statistical phenomenon back to the very end of the nineteenth century. He also notes that it took statisticians some time to accept the idea that *causal* considerations are a way of dealing with this so-called 'paradox'.

In the development of a probabilistic theory of causation Simpson's paradox enters as part of an argument by Nancy Cartwright against Humeanism. Cartwright (1983, pp. 23 f.) illustrates the problem with a hypothetical example about the relation between smoking (S) and the development of heart disease (H). It is a widely accepted fact that smoking may cause heart disease, i.e. that it raises the probability of developing heart disease:

$$P(H|S) > P(H|\sim S)\ . \tag{2.15}$$

It is perplexing though that it is possible to imagine circumstances in which this expectation is not met. Suppose, for the sake of argument, that smoking is correlated with some further factor that promotes health *and* that this further factor (statistically) outweighs the negative effects of smoking. One may suppose (just hypothetically) that smokers tend to exercise more, while non-smokers tend to exercise less. If one assumed that exercising prevented heart disease more effectively than non-smoking, it could seem as if smoking did not cause heart disease. This

[9] This work also proposes a more recent account of how to deal with Simpson's paradox.

could result in the conclusion that smoking had (apparently) no effect on heart disease

$$P(H|S) = P(H|{\sim}S) \tag{2.16}$$

or even the conclusion that smoking (apparently) prevented heart disease

$$P(H|S) < P(H|{\sim}S)\ . \tag{2.17}$$

The problem is that the Eq. (2.16) and the inequality (2.17) conditionalise on the total population of exercisers and non-exercisers. Conditionalising on the two sub-populations recovers the probability increase (Cartwright, 1983, p. 24). Let E be the sub-population of exercisers and ${\sim}E$ be non-exercisers. Then these two inequalities will hold:

$$P_E(H|S) > P_E(H|{\sim}S)\ , \tag{2.18}$$

$$P_{\sim E}(H|S) > P_{\sim E}(H|{\sim}S)\ . \tag{2.19}$$

The basic idea behind selecting these sub-populations is to hold fixed the factor *exercising* that gives rise to the spurious correlation. When it isn't taken into account, heart disease can be spuriously correlated with non-smoking (2.17). Partitioning the population into exercisers and non-exercisers is what fixes the context. In each subpopulation the factor is either set to exercising or non-exercising, which results in two situations in which the effect of smoking can be observed (as the non-spurious correlation between smoking and heart disease). However, in order to single out non-spurious correlations in real circumstances it won't suffice to fix a single factor, since there are usually a great number of factors that might interfere, which means that one has to take into account a long conjunction of situations in which such factors are held fixed (Cartwright, 1983, p. 26; Cartwright, 1989, pp. 55 f.).

Simpson's paradox is, as Cartwright argues, indicative of the need to supplement laws of association with causal laws. In their most basic form laws of association would be strict regularities, as thought of by Hume, but in this case they are of course probabilistic relations. The recovery of the probability increase in the inequalities (2.18) and (2.19) rests on *causal presuppositions* that guide the choice of the right population, i.e. the two sub-populations instead of the total population (Cartwright, 1983, p.24 f.). Elsewhere, Cartwright coins the slogan 'no causes in, no causes out' (Cartwright, 1989, Chap. 2). She postulates that the connection between causal laws and laws of association is as follows:

> 'C causes E' if and only if C increases the probability of E in every situation which is otherwise causally homogeneous with respect to E. (Cartwright, 1983, p. 25)

This condition is meant to specify the circumstances in which causal laws are recognisable and is followed by a technical exposition of 'causally homogeneous situations' (Cartwright, 1983, p. 26). Such situations as described by Cartwright are otherwise known as 'causal background contexts' (Eells, 1991, p. 85). The accompanying idea that a cause has to raise the probability of its effects in every background context is known as 'context unanimity' (Eells, 1991, p. 86). Whether this specific requirement is reasonable has been debated. Brian Skyrms suggested that it might suffice to demand that the effect raises the probability in at least one context without lowering it in others (Skyrms, 1980, pp. 108 f.). This proposal has the consequence that the causal concepts that result from such a theory differ from theories that commit themselves to Eells's context unanimity: Skyrms's theory groups causes that unanimously raise the probability of their effects together with causes that raise the probability of their effects only sometimes (and fail to do so on other occasions). Theories that accept context unanimity exclude causes that sometimes fail to raise the probability of their effects from being called 'positively relevant' or 'positive causes'. Ellery Eells argues that context-unanimity theories are superior in the sense that they 'have greater descriptive power' (Eells, 1991, p. 96). He states that such a theory can deliver various conceptual distinctions where K_i are background contexts:[10]

> X is a positive causal factor for Y in population P if and only if, for each i, $Pr(Y|K_i\&X) > Pr(Y|K_i\&{\sim}X)$. (Eells, 1991, p. 144)

Further concepts are easily obtained by substituting '>' with '<' for a negative causal factor and '=' for a neutral factor. In addition Eells also introduces the concept of a *mixed* cause, which is defined as being neither a positive, negative or neutral factor (Eells, 1991, p. 144). The most inclusive concept that completes the list is that of being 'causally relevant', which is a union of positive, negative and mixed causes (Eells, 1991, p. 106).

This way of dividing up causal concepts isn't universally accepted. Cartwright, for instance, doesn't introduce the concept of a mixed cause, which is inter alia reflected in her treatment of interactive causes, in contrast to the approach taken by Eells (see Sect. 3.3).

A related point of contention is whether contextual unanimity should be rejected altogether. John Dupré argued that the commitment to contextual unanimity is a leftover from a deterministic worldview that assumes 'that there is some finest level of description under which we can specify the complete causal truth about a situation' (Dupré, 1993, p. 204). He criticises Eells's position for relativising causal claims to specific populations, which needed to be specified to such a high degree that it became implausible that one should have any knowledge of them. The alternative suggested by Dupré is to look at the role of randomised controlled experiments. Dupré suggests that an experiment for answering the question of whether smoking causes heart disease would have to randomly select a group of infants, and separate them randomly into one group that is forced to smoke and a second group that is prevented from smoking, where for both groups the individuals'

[10] Eells proposes a more complicated account of background contexts, which he calls 'revised context approach' (Eells, 1991, p. 131). The way I present the position here is simplified.

lifespans are recorded. It is obvious that there are many unknown interfering factors at work in such an experimental design. Yet the reason why experiments are conducted this way is that it is assumed that the different distorting factors balance and possibly cancel out each other (Dupré, 1993, pp. 202 f.). Even though this particular example (viewed from a moral perspective) tends to undermine Dupré's argument that his approach is less of an 'epistemological embarrassment' (Dupré, 1993, p. 200), one has to acknowledge that pharmaceutical studies are designed similarly (except that they select adults who participate on a voluntary basis).

The role of experiments is also paramount in Cartwright's reasoning. Although Cartwright admits that probabilities 'can measure causes', she also maintains that this requires 'an impossible amount of information to begin with' (Cartwright, 1989, p. 62). Like Dupré, Cartwright discusses randomisation, but she also explores an experimental design (totally controlled experiment) that actually eliminates disturbing factors (Cartwright, 1989, pp. 62 ff.).

Contrary to Dupré and Cartwright's position are views such as those of Clark Glymour, Richard Scheines, Peter Spirtes and Kevin Kelly. These authors argue that it is a mistake to rely exclusively on controlled experiments, since many experiments can't be conducted for practical or ethical reasons. They suggest that so-called causal modelling has to fill this gap. Causal modelling aims to deduce causal relations purely from statistical data. Owing to the circumstance that such an approach involves quite a few computations, the concrete task of calculation is handed over to computer programs. An early example of such a program is TETRAD by Glymour et al. (1987).

Causal modelling is an axiomatic approach. This means the starting point is *not* to ask how the colloquial notion of cause is best captured by a definition. The approach is rather to set axioms and to argue that the theory that follows from such axioms provides notions of cause and effect that are close enough to what one is aiming for (Spirtes et al. 2000, p. 3).

In causal modelling, (presumed) causal relations can be graphically depicted as a causal graph. A causal graph is an ordered pair of a set of vertices and a set of edges (Pearl, 2009, p. 12).[11] The vertices (corresponding to variables) are referred to with terms denoting kinship, such as 'parents' or 'descendants' (Pearl, 2009, pp. 12 f.).

Whether a causal graph states a valid causal hypothesis is decided by referring to three axioms that constrict those graphs: Firstly, where the causal graph is a directed acyclic graph, i.e. one where all edges point in the same direction without forming any cycles, the Causal Markov Condition is imposed. The Causal Markov Condition is an axiom linked to Reichenbach's notion of screening-off. Let $\mathbf{V}$ be the set of variables that corresponds to the vertices of the graph and let P be a probability distribution over the variables in $\mathbf{V}$. Then the general idea is that in a directed acyclic graph P has to satisfy the Causal Markov Condition, which means that every variable X in $\mathbf{V}$ is screened off by its parents from all nondescendants of X

[11] Spirtes et al. (2000, p. 7) define a causal graph as an ordered triple with a set of marks as a third object. The marks (or arrowheads on the edges) are implicit in Pearl's account.

(Spirtes et al., 2000, pp. 29 f.; Pearl, 2009, p. 30). A second axiom is the minimality condition, which mandates that no proper subgraph is allowed to satisfy the Causal Markov Condition, i.e. that the hypothesised structure is minimal (Spirtes et al., 2000, pp. 30 f.). A third axiom is the faithfulness condition, which demands that there must *not* be any conditional independent relations in P that are not entailed by the Causal Markov Condition (Spirtes et al., 2000, p. 31).[12]

Proponents of causal modelling accounts are aware of the fact that their approach sometimes fails to retrieve the correct causal structure. Spirtes et al. (2000, pp. 38 ff.) admit, for instance, that the faithfulness condition doesn't always select the correct causal graph. The obstacle faced is yet again Simpson's paradox. If one takes Simpson's (1951, p. 241) example of a (fictitious) drug study where the effect of the treatment vanishes in the combined population but is recovered by conditionalising on female and male subpopulations, it turns out that the graph in which sex is caused by treatment gets erroneously selected. Spirtes, Glymour and Scheines brush this off by arguing that in the light of additional information, i.e. that the variable in question reports the sex of a person, they would prefer the graph in which sex isn't caused by treatment. Nevertheless, in the absence of such additional information they believe that it was admissible to apply the faithfulness condition anyway, since there was nothing to 'lose' (Spirtes et al., 2000, p. 41).

Cartwright objected to this that it was quite obvious that 'we lose getting the causal structure right' (Cartwright, 2007, p. 67). In the absence of further information there was no rational choice between the competing graphs: 'when you don't know, you don't know' (Cartwright, 2007, p. 68).

Another point on which Cartwright disagrees is whether the Causal Markov Condition is reasonable. While Cartwright agrees that holding fixed causal parents is in principle a sensible thought, she contests the notion that causal modelling succeeds in its application of the Causal Markov Condition and argues that there are various counterexamples (Cartwright, 2007, pp. 74 ff.).

However, the opposition between Cartwright and proponents of causal modelling is perhaps better understood when one takes into account how these authors view the world in general. Spirtes, Glymour and Scheines acknowledge that there are limits to the Causal Markov Condition such as reasoning in quantum mechanics where indeterminism poses a problem, but they insist that this 'is insufficient reason to abandon it in other contexts' (Spirtes et al., 2000, p. 37). Take for instance the above example of Salmon's interactive fork where we are confronted with an apparent failure of the common cause principle. Spirtes, Glymour and Scheines argue that this is no counterexample to the Causal Markov Condition, since there was reason to believe that a more fine-grained description of the case including the exact momentum of the cue ball and so on would restore 'conditional independence', i.e. the two effects (the drop of each ball into the pocket) would be screened off from one another (Spirtes et al., 2000, p. 37). Cartwright, by contrast, rejects the assumption that all non-quantum phenomena could be restored in a way that conforms with

[12] Pearl makes a similar assumption called 'stability' (Pearl, 2009, p. 48).

the Causal Markov Condition. While Cartwright doesn't claim that the macroscopic world behaves indeterministically, she does maintain that '[i]t is for the most part an unruly mess' (Cartwright, 1999, p. 110). She also rejects Spirtes et al.'s view (2000, p. 38) of how widely scientists apply the Causal Markov Condition. Cartwright argues that Spirtes, Glymour and Scheines overlook how heavily science relies on randomised controlled experiments (Cartwright, 1999, p. 113).

To sum up, probabilistic theories of causation can take different forms, since they are motivated from more than one perspective. Reichenbach tried to reduce time to causal order. Suppes took time as given, but thought that a probabilistic theory of causality was needed in order to overcome the ill-fitting notion of causation as necessitation. The debate between Eells and Dupré makes it apparent that a proponent of a probabilistic theory of causality need not reject determinism. The dispute between Cartwright and the proponents of causal modelling lays bare how theorising about probabilistic causality is intertwined with the issue of how causal knowledge is generated by science. A consequence of this is that these accounts result in competing notions of cause.

2.4 Interventionist Theory

Interventionist theory is an approach that is meant to complement the theories of Spirtes et al. (2000) and Pearl (2009). James Woodward, the major proponent of this theory, describes his account as a supplement that provides 'an account of the meaning or content of just those qualitative causal notions that Pearl (and perhaps Spirtes et al.) take as primitive' (Woodward, 2003, p. 38). Woodward thinks of his account as 'semantic or interpretive' and does not follow the aim of contributing 'to practical problems of causal inference' (Woodward, 2003, p. 38). Since Woodward's position is largely aligned with Spirtes, Glymour, Scheines and Pearl it cannot surprise that Woodward subscribes to the Causal Markov Condition (Woodward, 2003, pp. 64 f.).

Besides this, interventionist theory also shares some rough similarities with counterfactual theory: both approaches invoke the idea that an effect somehow depends on its cause. A difference is, however, that interventionism does not try to reduce causality to non-causal facts (the Humean mosaic). Its aim is to explicate interrelated causal terms to show how causal reasoning is interrelated with the practice of manipulating one's environment. Manipulation requires counterfactual knowledge: someone who desires to bring about a certain effect has to know what would happen if one were to change something in the environment. This gives the approach an anthropocentric flavour. However, proponents of interventionism such as James Woodward try to dispel this worry and claim that an interventionist theory can satisfy 'reasonable expectations about the objectivity of causal relationships' (Woodward, 2003, p. 12).

Interventionist theory is a type-level approach that tries to articulate type-level causal claims (Woodward, 2003, pp. 39 f.).[13] This is different from the Lewisian counterfactual account that articulates token-level causal claims. A further difference is in the way that causal relations are represented. Counterfactual theory takes the causal relata to be events. But interventionist theory uses variables instead. According to Woodward, 'variables are properties or magnitudes that [...] are capable of taking more than one value' (Woodward, 2003, p. 39). Woodward assumes that '[v]alues of variables are always possessed by or instantiated in particular individuals or units, as when a particular table has a mass of 10 kg' (Woodward, 2003, p. 39). He also suggests that one can use 'so-called indicator variables, taking the values 1 or 0, to express the occurrence or nonoccurrence of [...] properties or event types' (Woodward, 2003, p. 112). An example would be the 'indicator variable H that takes the values $\{hit, not\ hit\}$ or $\{1, 0\}$' (Woodward, 2003, p. 112). The introduction of indicator variables allows Woodward to paraphrase causal claims that are made in terms of 'event talk' rather than in terms of 'variable talk', i.e. properties or magnitudes.

This 'variable talk' has consequences for the language used in the account. Proponents of interventionism speak of 'causal relevance': instead of saying that 'X causes Y', it is more common to say that 'X is causally relevant to Y'. Although this is a linguistic side aspect, one should be aware of it. In contexts that lie outside interventionism 'causal relevance' can have a different meaning: 'causally relevant' is often used as a term that is opposed to 'causally productive'. In the chapters below I will have to use the term in both senses. So, it is important to know the difference: within interventionism 'causally relevant' is (roughly) synonymous with 'causes', but outside interventionism it can denote one specific sort of causal relation in contrast to another sort of causal relation.

In an interventionist account a variable is causally relevant to another variable in the case that an *intervention* on the *one* variable would change the value of the *other* variable (Woodward, 2003, pp. 39 f.).

The notion of an intervention is not an original idea of Woodward's. The idea occurs in Spirtes et al. (2000, pp. 299 & 308) and Pearl (2009, pp. 68 ff.). Woodward traces the idea even further back to the mid-twentieth century (Woodward, 2003, p. 39). In Woodward's account the notion of an intervention is at the heart of the theory, hence *interventionist* theory.

Woodward thinks 'of an intervention as an idealized experimental manipulation' (Woodward, 2003, p. 94). The aim of the account is to provide *truth conditions* for causal claims. Interventionism is *not* thought to be a practical guideline for experimentation (Woodward, 2003, p. 95).

But if it is not a practical guideline, why is it of philosophical interest to know these truth conditions? The answer to this theoretical question requires a practical

[13] Woodward also has an account of token-level causal claims, i.e. singular causal explanation (Woodward, 2003, pp. 209 ff.). But that account is not a central aspect of the theory. Woodward's main focus is clearly on type-level causation.

example. Let us consider another pharmaceutical study, in this case one that tries to answer the question of whether administering a certain drug helps patients to recover from a disease. Woodward (2003, pp. 94 f.) gives the same example. Please note that I shall strip off the 'variable talk' for the sake of clarity.

If one were to take a *naive* approach, one would simply administer the drug and wait and see. (One may speculate whether ancient medical knowledge was actually acquired that way: administer a herbal infusion and wait and see.) However, it is common knowledge that belief in recovery can help patients to recover from a disease: the so-called placebo effect. If patients have a strong belief in the efficacy of a drug, they can recover more quickly, *even though* the drug in itself may have no effect or very little effect. The naive wait-and-see approach has difficulty telling these effects apart. The way out is to employ a more sophisticated approach, i.e. a controlled and randomised experiment (as already mentioned in Sect. 2.3). One has to study two randomly selected groups: one group receives the drug, the other group receives the placebo only. If the participants don't know to which group they belong, the approach can statistically tell apart the two effects, i.e. the placebo effect and the effect of the drug.

In *practice* there is no choice but to resort to a comparison with a control group in order to distinguish the two effects. In *theory*, however, the problem can be solved differently: one would have to distinguish manipulations that have unintended side effects from manipulations that have an isolated effect. In practice one cannot administer a drug without also initiating the process that makes patients believe in their own recovery (the placebo effect). But in theory it would of course suffice to know that a manipulation has an *isolated* effect on the physiology and not the psyche. This sort of manipulation, which precludes side effects or any other correlation that would obscure what the manipulation does, is called an 'intervention'.

The objective of proponents of interventionism is to give a theoretical account of that sort of manipulation. An exhaustive account of an intervention is given in Woodward (2003, Chap. 3). A further interest followed by Woodward is the explication of interrelated causal concepts rather than a single causal concept. The account makes distinctions such as those between direct, total and contributing cause (Woodward, 2003, pp. 45–61). Woodward maintains that 'manipulation is a vague idea that needs to be made more precise along many different dimensions' (Woodward, 2003, p. 45). He interprets this task as '*revisionary* or *normative*' (Woodward, 2003, p. 7, italics in original), which suggests it is a parallel to the axiomatic approach in causal modelling accounts.

This is, for instance, reflected in his thesis that any causal relation has to be reproducible. Woodward argues that such a requirement is needed for token- as well as type-level claims, if we want to rule out 'the possibility that there are true type-causal claims that are singularist in content in the sense that it is built into their meaning that they have implications only for what happens in a single instance or a one-shot causal interaction' (Woodward, 2003, p. 72). The problem is not that there could be contingently singularist causal claims with only one instance, but that there could be *conceptually* singularist causal claims: in Woodward's opinion it wouldn't be a problem to have a type-level causal claim with only one instance, but it would be problem if it was for conceptual reasons impossible to have more than one instance.

In other respects the theory is less demanding. As I have said above, its aim is not to reduce causality. The notion of an intervention is a *causal* notion already. What it means 'to intervene' is only explicable in terms of causal concepts. He explicates the notion of an intervention as a special sort of manipulation that has an isolated effect. Both 'manipulation' and 'effect' are causal concepts. Woodward, too, doesn't provide any reductive account in terms of non-causal concepts. Hence, Woodward's definitions of causal concepts must be circular. Consider, for instance, the definition of a direct cause:

> (**DC**) A necessary and sufficient condition for X to be a direct cause of Y with respect to some variable set $\mathbf{V}$ is that there be a possible intervention on X that will change Y (or the probability distribution of Y) when all other variables in $\mathbf{V}$ except for X and Y are held fixed at some value by interventions. (Woodward, 2003, p. 55)

This definition relies on the notion of an intervention. Woodward thinks, however, that it is non-vicious circularity. Although one has to explicate an intervention in causal terms, the definition still leaves open the question of whether there is a causal relationship between the variables X and Y (Woodward, 2003, pp. 104 f.).

Another difficulty with the above definition is that it uses modal terminology. It doesn't say that there has to be an intervention, it only demands that an intervention is *possible*. This retreat into modal terminology is made in order to avoid anthropocentrism. The definition is thought to cover those interventions, too, that are not in fact possible for human beings. However, the word 'possible' brings with it a burden: it requires an underlying notion of modality that has to be made precise. Where it is unclear how to interpret this modality, the definition itself is unclear.

Nevertheless, if one accepts Woodward's approach, other interesting concepts are available. The notion of a direct cause is accompanied by two further concepts of cause that allow for better differentiation between causal claims. The notion of a contributing cause

> is intended to capture the intuitive idea of X influencing Y along some route even if, because of cancellation, X has no total effect on Y (Woodward, 2003, p. 50).

The notion of a total cause says that

> (**TC**) X is a total cause of Y if and only if there is a possible intervention on X that will change Y or the probability distribution of Y (Woodward, 2003, p. 51).

The same ideas are referred to as 'component' and 'net effect' by Hitchcock, who notes that his and Woodward's definitions are virtually equivalent (Hitchcock, 2007, p. 208). These concepts can help to disentangle seemingly contradictory causal claims such as remarked on by Germund Hesslow (1976), an example discussed by Woodward (2003, pp. 49 f.). Hesslow noticed that Suppes's probabilistic theory of causation (Sect. 2.3) would return contradictory causal judgements in certain cases. His example was to consider the effect of contraceptive pills. These have the known side-effect of causing thrombosis. However, thrombosis is also caused by being pregnant. Hence, it would appear that this contraceptive can also prevent thrombosis by virtue of preventing pregnancies. This apparent contradiction is, however, easily resolved once one distinguishes between the average effect in the total population (net effect/total cause) and the effects in the subpopulations of women who are pregnant and women who use the contraceptive (component effect/contributing cause).

To sum up, the interventionist theory is a non-reductive account of causation that explicates how causal claims are related to the practice of manipulation. It is a type-level theory and it supposes that the causal relata are properties or magnitudes instead of events. An intervention is a special sort of manipulation that can be used as a means of analysing causal relations. Furthermore, interventionist theory is driven by revisionist aspirations, among them the idea that all causal claims (type and token) have to accord with the reproducibility requirement. Nevertheless, the concepts of cause and effect obtained from the theory can help to disentangle seemingly contradictory claims.

2.5 Process Theory

Process theory is the account that was developed by Wesley Salmon once he had given up on pursuing a probabilistic theory of causality (Sect. 2.3). It represents a further response to the problems that were raised by David Hume. However, the solution he offers is not the same as in the traditional approaches. Salmon claims that it is, *pace* Hume, possible to identify 'the causal connections that Hume sought and failed to find' (Salmon, 2002, p. 28). He urges us to 'take Hume's challenge seriously' and to 'try to find a physical connection between cause and effect' (Salmon, 1998, p. 16). This, he says, may not have been possible at Hume's time but it 'seems possible in the twentieth century' (Salmon, 1998, p. 16). As a caveat it should be added that Salmon does not give up on the Humean *empiricist* programme. Salmon makes it clear that he doesn't want to 'introduce any "occult" concepts of "power" or "necessary connection"' (Salmon, 1998, p. 113). He presents his account of causation as a *realist* one, which removes the concept of causality 'from the field of metaphysics and transports it to physics' (Salmon, 2018, p. 131).

Salmon's thesis is that the physical connection between cause and effect is a process. He uses Russell's notion of a causal line as a starting point for his account of a process. Russell defines a 'causal line' as

> a temporal series of events so related that, given some of them, something can be inferred about the others whatever may be happening elsewhere. A causal line may always be regarded as the persistence of something—a person, a table, a photon, or what not. Throughout a given causal line, there may be constancy of quality, constancy of structure,

or gradual change in either, but not sudden change of any considerable magnitude. (Russell, 1948, p. 477)[14]

As Salmon notes, this definition is in accordance with Hume's account of causal relations: it accommodates 'contiguity, priority, and constant conjunction' (Salmon, 1998, p. 196). However, in Salmon's opinion Russell fails to notice that his definition is too inclusive. It applies to causal but also to non-causal processes. Non-causal processes, which are called 'pseudo-processes' by Salmon, are distinguishable from causal processes by their inability to transmit a mark. Mark transmission is a criterion overlooked by Russell.

Another way of saying that a process can transmit a mark is to say that it can transmit information (Salmon, 1998, p. 114). The ability to transmit information is a crucial feature of causal processes. Salmon points out that, '[a]ccording to the theory of relativity, it is not possible to send a signal at a velocity greater than the velocity of light in vacuum' (Salmon, 1998, p. 16). One may use this knowledge as a criterion for distinguishing causal processes from non-causal processes: causal processes are incapable of travelling faster than the speed of light but they are capable of transmitting information; non-causal processes are capable of travelling faster than the speed of light but they are incapable of transmitting information (Salmon, 1998, p. 16). Salmon emphasises that mark transmission isn't 'any mysterious kind of necessary connection' (Salmon, 1998, p. 119). He instead thinks of it as 'a species of constant conjunction' (Salmon, 1998, p. 119).

The difference between the two kinds of process can be illustrated by means of the following example: Suppose that there is a rotating beacon on top of a lighthouse in the centre of a circular room, or arena, or some kind of screen that runs around the beacon. As the beacon rotates, there are two processes going on. One process is the light travelling from the beacon to the screen. The other process is an illuminated spot that travels over the surface of the screen. The former process is a causal process. It is possible to demonstrate this by marking the process. It is possible to interpose a red filter, for instance. The light, as it travels through the filter, becomes red and it retains the red colour beyond the place of the marking. One can use such a process as a means of communication. If the sender (the one who puts a filter into the light) and the receiver (the one who watches the light) have agreed on some convention of communication, they can transmit a message. One could agree, for instance, that turning the light red means 'Rudolf the red nosed reindeer has just arrived' or something like that. The same is not possible with the process that runs in a circle over the screen. If I am standing at the screen and try to mark the process by holding up the red filter, I will not succeed in passing on a message along that process. I may turn it temporarily red at the place where I stand, but the colour will not be retained. The colour turns back to white the moment the luminous spot travels away from me. It is impossible to mark this process. Hence, it is a pseudo-process. A further indication of its non-causal nature is that it can travel at a superluminal velocity. If one enlarges the radius of the screen to a certain extent, the luminous spot

[14] Also quoted in Salmon (1998, p. 196) and partially quoted in Salmon (2002, p. 29).

will eventually travel faster than the speed of light (Salmon, 1998, pp. 16 f./194 f.; Salmon, 2002, pp. 29 ff.).[15]

The pseudo-process running around in a circle exhibits all the characteristics of a 'causal line': it is a temporal sequence of events, the luminous spot occurs in one place and then in a neighbouring region, and the process exhibits a regular behaviour that allows us to infer where the luminous spot will occur in the next moment. Furthermore, the luminous spot exhibits persistence, i.e. it doesn't cease to exist while it travels over the screen. It also has a constancy of quality and structure. Hence, one has to agree with Salmon that some further requirement, such as mark transmission, should be introduced in order to distinguish non-causal from causal processes.

Salmon also distinguishes causal and non-causal intersections between processes. A spatiotemporal intersection between processes occurs '[w]hen two processes, causal or pseudo-, meet each other' (Salmon, 1998, p. 17). A causal intersection is called 'causal interaction' (Salmon, 1998, p. 17). A causal interaction requires that both processes get modified and that 'the modifications persist beyond the place of intersection' (Salmon, 1998, p. 17). If there are no modifications that are retained, there is a non-causal intersection only (Salmon, 1998, p. 17).

Salmon distinguishes three types of interaction between causal processes: X-type, Y-type and λ-type. Salmon gives paradigm examples for each type of interaction. A collision of balls is an example of an X-type interaction: the balls approach each other, they collide, and they continue their way in different directions. Two processes enter and two processes leave the place of the interaction (the collision). A hen laying an egg is an example of a Y-type interaction: there is the hen, there is the hen laying the egg, and then there are the hen and the egg. One process enters and two processes leave the place of the interaction (the laying of the egg). A snake eating a mouse is an example of a λ-type interaction: there is the snake, there is the mouse, and then there is the mouse inside the snake. There are two processes that enter and one process that leaves the place of the interaction (the eating of the mouse) (Salmon, 2002, pp. 31 f.).[16]

However, a problem with this classification of causal interactions is that the mark-transmission criterion applies to X-type interactions only (Salmon, 2002, p. 32). At least, it is not clear what happens to a mark in Y-type and λ-type interactions: is the mark created anew or is it consumed in the interaction?

In addition to this there is a further problem: Salmon is particularly worried that he may have to rely on counterfactual conditionals. In the lighthouse case there is, for instance, the problem that one can interfere in such a way that it would seem as if the pseudo-process transmits a mark. The spot of light as it travels over the surface of the circular screen is a pseudo-process since it cannot retain a mark beyond the place

[15] Salmon repeats this example in various places and versions. I have repeated it with some minor modifications of my own.

[16] Note that the letters 'X', 'Y', and 'λ' are supposed to be 'read' from bottom to top: they visualise the geometrical shape of the causal interactions.

of modification. However, imagine the following: As before, I am standing at the screen and I hold up a red filter. The spot turns red but it regains its white colour as it travels on. Now imagine that at the time when the spot reaches the red filter someone *also* replaces the white lens at the beacon with a red lens. In that case the spot that travels around the screen would not return to its original colour. The spot would remain red instead and '[i]t would seem that the luminous spot had transmitted a mark' (Salmon, 1998). A solution would be to introduce a counterfactual condition. In the counterfactual case that no one replaces the lens at the beacon at the same time that I hold up the filter, the process would not retain the mark. However, Salmon doesn't want to introduce counterfactual conditionals into his analysis. They are in his opinion not well enough understood (Salmon, 1998, p. 18). Instead, his response is to introduce conserved quantities into his theory (Salmon, 1998, pp. 18 ff.), which is an idea that he adopts from Phil Dowe (1992).

With this modification it is no longer central to ask whether a process has the ability to transmit a mark but whether it does in fact transmit a conserved quantity such as 'linear momentum, energy, or electric charge' (Salmon, 1998, p. 18). A conserved quantity is a quantity that is subject to a conservation law. According to the first law of thermodynamics, for instance, energy is neither destroyed nor created (Salmon, 1998, p. 19).

The moving spot in the above example does not transmit a conserved quantity. At the place where the spot is reflected there is of course some conserved quantity present (since light is a sort of energy), but it is not *transmitted* along the process. So, one can rule out the moving spot as a causal process. Salmon prefers this analysis, since it 'does not involve counterfactual propositions' (Salmon, 1998, p. 19).

Furthermore, it is also easier to deal with Y-type and λ-type interactions: in a Y-type interaction such as a hen laying an egg conserved quantities are divided between two processes, while in a λ-type interaction conserved quantities are merged in a single process. The crucial point is that there is some exchange of a conserved quantity in the intersection of processes (Salmon, 2002, p. 33).

But although Salmon abandons mark transmission as the criterion that distinguishes causal from non-causal processes, he does still insist 'that the mark method is an *extremely useful* tool for the discovery and study of causal processes' (Salmon, 1998, p. 20). An important application of the method is, for instance, 'the use of radioactive tracers in studying physiological processes' (Salmon, 1998, p. 20).

A process theory that is formulated in terms of conserved quantities can be summarised using just two propositions, as Dowe demonstrates:

> CQ1. A *causal process* is a world line of an object that possesses a conserved quantity.
> CQ2. A *causal interaction* is an intersection of world lines that involves exchange of a conserved quantity. (Dowe, 2000, p. 90, italics in original)

To speak of a 'world line' is practically the same as speaking of a 'process'. If summed up like this, process theory turns out to be a plain theory: (at first sight) there are no complicated extra conditions. However, simplicity has its price. Dowe is forced to conclude that omissions can be neither causes nor effects. This leads him into a conceptual distinction between 'causation' and 'causation*' (Dowe,

2000, p. 123). Another term used instead of 'causation*' is 'quasi-causation' (Dowe, 2001, p. 217). The only reason for introducing the concept of 'quasi-causation' is to give an account of the practice of speaking of omissions as (quasi-)causes or (quasi-)effects.

To sum up, process theory is another response to Hume's challenge. Process theory is an approach that is in accordance with the ideas that causal relations are contiguous in time and space, that the cause has temporal priority to the effect, and that causation is a sort of regular succession. However, while regularity theory and counterfactual theory allow omissions or absent factors to be spoken of as causes and effects, process theory instead forces the concept of quasi-causation to be introduced.

2.6 Normality Theory

The approaches to causation that I subsume under the label of 'normality theory' are those that seek to illuminate an inegalitarian notion of cause. An important example to which the label 'normal' can quite naturally be attached is H. L. A. Hart and Tony Honoré's (1985) account of causation. These authors deny what I have called the 'Millian arbitrary selection principle' (in Sect. 2.1) and argue that conditions are no causes. They think that 'it is plain that our choice, though responsive to the varying context of the particular occasions, *is not arbitrary or haphazard*' (Hart & Honoré, 1985, p. 11, italics added). I read this as the claim that objective choices are possible. However, not everyone reads Hart and Honoré that way, i.e. with the emphasis on those arguments that find objective criteria for singling out causes in contrast to their conditions. Notably, Mackie (1980, Chap. 5) reads Hart and Honoré in a way that focusses on how interest may compromise the objectivity of causal judgements. It is true that Hart and Honoré's arguments are hedged about with many cautious remarks and caveats. Yet we should avoid being distracted from their thesis that it is *wrong* to claim that 'every factor necessary for the occurrence of an event is equally entitled to be called "the cause"' (Hart & Honoré, 1985, p. 12). They defend this view with an analysis of common-sense beliefs from which they conclude that there is a central notion of cause that is based on an intuitive understanding of normality:

> The notion, that a cause is essentially something which interferes with or intervenes in the course of events which would normally take place, is central to the common-sense concept of cause, and at least as essential as the notions of invariable or constant sequence so much stressed by Mill and Hume. (Hart & Honoré, 1985, p. 29)

This position is in stark contrast to such views as Lewis's counterfactual theory of causation that aims at an articulation of the notion of *a* cause instead of *the* cause (Sect. 2.2). Other positions such as Mackie's regularity account seem to lie in between. Mackie tries to satisfy both egalitarian and inegalitarian intuitions: the inus condition does justice to the Millian arbitrary selection principle, but the idea of postulating a causal field is motivated by the assumption that some circumstances are *normal* and others are not (Sect. 2.1). One might therefore argue that the distinction between egalitarian and inegalitarian accounts of causation is not always clear cut and that it can be an oversimplification to put accounts in either one category or the other.

Those who believe that a distinction between causes and conditions must always be a matter of convention may point out that Hart and Honoré's subject matter is legal theory, where they are concerned with guilt and liability, which makes it in turn plausible for them to investigate an *in*egalitarian rather than an egalitarian concept of cause: a culprit is often a person who has done something which is considered to be *the* cause of harm or damage. When it comes to undisturbed nature, however, some may think that one should not invoke the idea of normality.

Yet even if one took such a view, one should nevertheless consider that causal intuitions about causes without human intervention can still be shaped by assumptions about what is normal. Hart and Honoré argue that an idea of normality is at least a partial guide to common-sense beliefs about how causes operate in the absence of humans:

> Analogies with the interference by human beings with the natural course of events in part control, even in cases where there is literally no human intervention, what is to be identified as the cause of some occurrence; the cause, though not a literal intervention, is a *difference* from the normal course which accounts for the difference in the outcome. (Hart & Honoré, 1985, p. 29, italics in original)

One could admit this as a mere hypothesis about the human psyche and still resist the thesis that the outcome of such human reasoning is objective. This approach, however, is disputed by Hart and Honoré. They distinguish the context from normal conditions and argue that the context determines the normal conditions. While the same type of occurrence might be considered a condition or a cause in *varying contexts*, it is not—according to Hart and Honoré—a cause or a condition in the *same context*. The absence or presence of oxygen, for instance, can be both normal and abnormal, depending on the context. When a fire breaks out in the open atmosphere of the Earth, oxygen is a condition, since it is normal for the Earth's atmosphere to contain oxygen. However, in an—as Hart and Honoré would see it—objectively different context, oxygen can also be a cause of a fire. In a factory where oxygen is normally kept away from certain production processes the presence of oxygen would be abnormal and therefore considered to be the cause (Hart & Honoré, 1985, p. 35).

Although the selection of normal conditions in a given context can be straightforward and objective, Hart and Honoré agree that there are also cases where the selection of normal conditions can be driven by interest rather than objective circumstances. They consider a hypothetical example of a famine and note that different perspectives can result in a selection of different normal conditions in one and the same context: the peasant attributes it to the drought, but the World Food authority[17] attributes it to the government's failure to store enough food (Hart & Honoré, 1985, p. 35).

At first glance this appears to corroborate the Millian thesis that one can arbitrarily select factors and call them 'causes'. Hart and Honoré maintain, however, that the peasant's hypothesis and the World Food authority's are *not* equally valid. The peasant who attributes the famine's cause to the drought has a limited view that doesn't take into account that there are 'customary techniques, procedures, and routines to counteract' the effects of the drought (Hart & Honoré, 1985, p. 37). The World Food authority, by contrast, has a broader perspective that includes not just the knowledge of the drought but also the knowledge of how governments can as a matter of fact (and not just hypothetically) prevent a famine. In other words, the selection of alternative normal conditions in one and the same context doesn't need to be interpreted as arbitrary. One can explain this if one takes into account that the peasant and the World Food authority are not equally well informed. As a consequence one should then accept the World Food authority's causal judgement since it takes more of the objective circumstances into account (Hart & Honoré, 1985, p. 37).

Another observation that Hart and Honoré make is that time is *no* criterion for deciding between the cause and its conditions. They argue that '[c]onditions are *mere* conditions as distinct from causes because they are normal and not because they are contemporaneous with the cause' (Hart & Honoré, 1985, p. 39, italics in original). A gentle breeze, for instance, can be a normal condition that follows upon the cause. If someone lights a fire and a breeze spreads the fire to another's property, it is the lighting of the fire that has caused the destruction of that property and not the gentle breeze, which occurred as *normal*, although subsequent to the cause (Hart & Honoré, 1985, p. 39).

However, all that Hart and Honoré present are paradigm examples which illustrate how one applies the idea of normality but don't give any technical definition of it. This means that even if one accepts their analysis of how common sense distinguishes conditions from causes, there is still the question of how to give a concrete definition of normality.

Gerhard Schurz (2001) argues, although not as a response to Hart and Honoré, that generalised evolutionary theory is the theoretical framework that allows one to develop an account of what 'normal' means. He examines how the notions of prototypical and statistical normality are related. Statistical normality is the idea of statistical majority, i.e. a trait is normal when it is in the majority. The criterion for prototypical normality, in contrast, is that a prototype possesses that trait. Given these two criteria, it is in principle conceivable that the two notions need not be congruent. Schurz considers a hypothetical scenario in which some environmental catastrophe results in a situation in which birds no longer use their wings for flying. Applying the prototypical notion of normal, one could then argue that it was still normal for birds to fly—despite the fact that the majority of birds stopped flying. In

[17] This is the way Hart and Honoré capitalise 'World Food authority'.

the light of such an example one might think of prototypical and statistical normality as disparate notions of normality. Schurz, however, maintains that although it is impossible to reduce prototypical normality to statistical normality, it is still true that prototypical normality (properly understood) implies statistical normality (Schurz, 2001, p. 478).

His argument relies on evolutionary selection as a connecting principle. In the special case of biological evolution, the argument proceeds by pointing out that the norm states of biological self-regulatory systems, i.e. traits of organisms, are the outcome of natural selection. Since natural selection accumulates fitness-enhancing traits, those traits or prototypical norm states had to be 'in the high *statistical majority* of cases and time points' (Schurz, 2001, p. 480, italics in original).[18]

Schurz argues that one can generalise this argument. Governments that cause economic decline are removed from office, which means that governments that keep the economy running are in the majority. Another of Schurz's examples is consumer electronics. If devices are unsafe, they lose market share. Such selection mechanisms connect prototypical normality with statistical normality (Schurz, 2001, p. 480).

This account, although not specifically developed as a reaction to Hart and Honoré, appears to be applicable to the above examples, too. One can argue that the absence of oxygen in the factory is normal since there is selection against factories in which oxygen is not kept away from such a production process. Likewise, there is selection against governments that are not taking precautions against famine. What is not covered by this generalised-evolutionary-theory approach is the question of why oxygen is a condition if a fire breaks out in the atmosphere. It is not obvious that any selection process is at work in this case.

The important point in this case is to recognise that a selection process is no precondition. Schurz's account demonstrates how to connect prototypical with statistical normality. However, the connection principle is only needed to account for the relation between prototypical and statistical majority. Where the statistical majority is known anyway, thoughts about a prototype may be expendable. It is obvious that on the Earth's surface oxygen is present in the overwhelming majority of situations, except for some hollows in which CO_2 from underground sources accumulates. Even the peasant's view would be explicable provided that the region one is talking about isn't a place where droughts are in the statistical majority. However, 'explicable' doesn't mean that the peasant is right in his or her assessment. The World Food authority would still be right to point out that the famine is caused by the government's mismanagement, since governments which counteract a famine are in the statistical majority. Whether one's causal judgement is correct or not

[18] Note that Schurz presupposes an extended time frame that isn't limited to the species but also includes all ancestor species in which a trait existed. This is meant to ensure that a change of function of a trait does not result in the conclusion that a trait has lost its proper function (Schurz, 2001, pp. 489 f.).

depends on facts. Incomplete knowledge of these facts can result in an error of judgement.

That the statistics have to be exhaustive is one of the aspects for which Schurz has been criticised. Wolfgang Spohn (2014, pp. 389 f.) argues that the required statistics are often not available. His alternative proposal is to develop a ranking theory for conditional beliefs that need not be evaluated as true or false. However, the fact that one will sometimes have to suspend judgement is not a reason to discard Schurz's position. That is not a deficiency but a merit of any realist theory. It would be a problem if Schurz's theory required a lot of in-principle claims that cannot be tested. But I don't see that this is the case. There is a clear conception of what needs to be known and in many cases it is in fact possible to acquire such knowledge by empirical investigation.

What could be a problem is that Schurz's account may not be suitable as an extension for Hart and Honoré's normality theory, due to concerns about circularity: the selection process that connects prototypical to statistical normality is itself a causal process. However, I don't think that this is detrimental. The connection principle explains why it is reasonable to expect that prototypical normality will usually imply statistical normality. But, as I have already indicated, Hart and Honoré's account can proceed with statistical normality alone. Hence, the circularity in question is non-vicious, similar to the non-vicious circularity that occurs in interventionist theory (Sect. 2.4).

An interesting position is taken by Kenneth Waters (2007), who reads Hart and Honoré from the same perspective as Mackie, i.e. emphasising caveats, while developing his own argument against Mill's arbitrary selection principle (which he refers to as 'causal parity'). Waters argues that there are objective differences between potential and actual difference-making causes. Potential difference-making causes are factors that are otherwise known as conditions. Waters's notion of actual difference-making causes is similar to Hart and Honoré's central notion of cause in that he ranks factors according to their importance or contribution, but his account differs in that he argues that there can be more than one cause that stands out against conditions, i.e. possible difference-making causes.

Waters's major claim is that those who uphold the causal-parity thesis, i.e. Mill's arbitrary selection principle, are mistaken in suggesting that any one context is as good as any other. Classical genetics serves as an example. Waters argues, contrary to what is sometimes suggested, that it is false to attribute to adherents of the Morgan school of classical genetics the belief that traits are caused by single genes. Although there are contexts in which single genes make the difference between the wild-type allele and its mutated form, Morgan and researchers in that tradition would not have deduced from this that there cannot be other causes of the phenotype. Waters suggests that they implicitly followed a rule, which he calls the 'difference principle', i.e. that 'differences in a gene cause uniform phenotypic differences in particular genetic and environmental contexts' (Waters, 2007, p. 558).

This rule, however, can only be complied with if researchers deliberately choose experimental conditions that are standardised in an appropriate way. That researchers followed this practice undermined 'the notion that they naively believed that each of the phenotypic characters they studied were preformed in single genes or even gene complexes' (Waters, 2007, p. 558). In other words, the controlled experimental setup is only intelligible if it is clear from the beginning that other factors affect the trait in question, too.

The example of a breeding experiment is meant to illustrate what it means to be *the* actual difference-maker. A prerequisite for there being the actual difference maker is that there are actual differences in the population. Or, more specifically, there needs to be actual variation in the effect—eye colour, for instance. If all this variation can be accounted for by the allele states in a single gene, then this gene is in that particular population *the* actual difference-making cause.

This account allows Waters to deal with problems similar to those considered by Hart and Honoré. Waters argues that striking a match is the actual difference-making cause in the actual population of matches of which one gets struck and the others lie dormant in the box:

> If Mary took the match from a box, then the matches contained in the box were an actual population and the difference between having lit versus not having lit was an actual difference. With respect to this actual effect (that is, the actual difference in this actual population), it is obvious that Mary's striking the match was the actual difference making cause and that the presence of oxygen was merely a cause. The presence of oxygen was not the cause that made the actual difference in this actual population of matches because, after all, oxygen was present for all the matches. (Waters, 2007, pp. 568 f.)

The parallel to Hart and Honoré's account is remarkable. Hart and Honoré would have reasoned virtually the same way, except that they would have said that the presence of oxygen was normal and that it was also normal for stored matches not to burn. What is different is that Hart and Honoré are less confident about stating this unconditionally. Waters, in contrast, claims

> that the subset of causes that are the actual difference makers of such an effect is fixed entirely by the ontology of causation in the actual population, not by a mix of ontology and interests as recent writers have maintained. (Waters, 2007, p. 569)[19]

There are also cases in which an actual difference in a population is not accounted for by a single factor but by a combination of factors which are actually different, and each of which partially contributes to the observed difference. Where this is the case, Waters speaks of such a factor or variable as '*an* actual difference maker' as opposed to '*the* actual difference maker' (Waters, 2007, p. 571, italics in original). This means that Waters's account distinguishes three categories: possible difference-makers (which others call conditions), actual difference-makers (which are elsewhere rarely singled out separately) and the actual difference-maker (which corresponds to Hart and Honoré's notion of cause).

A further account that follows Hart and Honoré's general claim that causes are abnormal occurrences is Andreas Hüttemann's (2013) disposition-based process theory. Central to that account is the analysis of part-whole explanations in physics. Such explanations proceed by decomposing larger systems into subsystems. The

[19] Waters is referring particularly to Jonathan Schaffer (2005).

behaviour of the combined system is then explained by showing how each subsystem contributes to the overall behaviour of that system (Hüttemann, 2013, pp. 105 ff.). What may seem a vague idea at first glance is in fact guided by laws of composition that allow a quantitative and not just a qualitative assessment of the contributions of the subsystems. An example of a simple compound system would be a stone falling through a medium, where one can then calculate how the medium affects the travel of the stone to the ground (Hüttemann, 2013, p. 109).

Hüttemann argues that this practice requires the attribution of dispositional properties. A dispositional property in Hüttemann's sense 'is a property that, if instantiated by an object, is manifest under specific conditions only' (Hüttemann, 2013, p. 107). Categorical properties are treated as limiting cases, i.e. properties that, if instantiated, are always manifest (Hüttemann, 2013, p. 107). Both the stone and the medium are disposed to behave in a certain way. The behaviour of each subsystem is, however, rarely or never completely manifest. There is usually a gradual manifestation of dispositions. In the case of the falling stone 'the medium serves as a (partial) antidote' to its disposition to fall to the ground (Hüttemann, 2013, p. 107). That those part-whole explanations are successful becomes intelligible as soon as one adopts the hypothesis that those subsystems possess a default behaviour, i.e. a dispositional property that shows itself if the subsystem is undisturbed. Hüttemann calls the laws that govern the default behaviour of the subsystems 'quasi inertial laws', analogous to Newton's first law (Hüttemann, 2013, p. 108).

Although this is an account of how part-whole explanations work, it is not yet an account of causation. An important detail is that the part-whole relation is atemporal in the sense that the contribution of the dispositions is simultaneous. This is not in accordance with the widely shared assumption that causation has a causal order with causes preceding their effects in time. A further complication is that one cannot solve this problem by accepting that simultaneous causation is possible. It would then still be a problem that the contributions of the dispositions in a part-whole explanation are not separable as distinct events, which clashes with another common assumption about causes and effects. Hüttemann concludes therefore that the contributions the dispositions make cannot be causes (Hüttemann, 2013, pp. 110 f.).

If dispositions, as they are understood in this account, are not causes, what can be a cause in this account? Hüttemann argues that disturbing factors can be a cause, since they still meet the usual criteria for causes. He uses billiard balls as an example. When balls on a billiard table bounce into each other there is a clear temporal order of events: the billiard balls first approach each other on linear paths, they then collide, and after that they continue to move on changed courses. The collision is a disturbing factor that acts as an antidote to the default behaviour of rolling in a straight line over the table (Hüttemann, 2013, p. 113).

Naturally, this example relies on idealisation. Hüttemann notes that the roughness of the table can also disturb the course of the billiard balls. When those disturbances are minor, Hüttemann suggests relegating them to a causal field, as is done in

Mackie's regularity theory. Having adopted that idea, Hüttemann provides this definition of a cause:

> Relative to a causal field, a cause is an actual disturbing factor (antidote) to the default behaviour that a system is disposed to display. (Hüttemann, 2013, p. 116)

This account, which focusses on actual disturbances of inertial behaviour, allows Hüttemann to avoid some difficulties which are faced by other theories. One problem that is easily avoided is counterexamples of preemption (Sect. 2.2). Those cases occur when a backup cause nullifies counterfactual dependence. Hüttemann's theory doesn't run into this problem since causes are always conceived of as actual disturbing factors and not hypothetical backup causes (Hüttemann, 2013, pp. 114 f.).

Another benefit is that a disposition-based process theory doesn't require the exchange of marks or conserved quantities. Such exchanges bring with them the problem of ersatz concepts of cause such as 'quasi-causation' (Sect. 2.5). Hüttemann, by contrast, characterises causal processes in terms of dispositions of systems. This means that gaps in which mark transmission or the exchange of conserved quantities are impossible are not detrimental to the theory. Hüttemann even goes so far as to suggest that default processes need not be spelled out in terms of physics. Other sciences, such as biology or economics, may characterise default processes in their own terms, provided those processes are concerned with the temporal evolution of a system. Likewise, the notion of disturbance can be specified differently in the respective sciences (Hüttemann, 2013, pp. 118 f.).

The commonality of the theories discussed in this section is that they articulate the notion of cause that is central to common-sense reasoning. All three approaches contain an element of context sensitivity. This, however, should not be confused with subjectivity. A key thesis in these accounts is that once the facts (e.g. the population in Waters's account) are settled, it is possible to make objective causal judgements relative to those facts. An important difference is how these accounts determine what—in Hüttemann's terminology—the dominant cause is. Hart and Honoré rely on an intuitive understanding of normality, which, as I have argued, can be supplemented with Schurz's account of prototypical and statistical normality. Waters avoids the question of what is normal by focussing on actual differences in particular populations, which is an approach similar to Hart and Honoré's position. The theory that differs most from the others is Hüttemann's disposition-based process theory. That approach doesn't need to answer what is normal, nor does it need to care about populations. The key element in his account is the notion of an inertial system, i.e. an idea that is quite well understood in physics, where default behaviour is routinely ascribed to subsystems of compound systems. What is interesting about this latter account is that it elucidates the common-sense concept of cause from a quite different perspective than that adopted by Hart and Honoré, or Waters. However, since all three accounts converge towards a similar notion of cause, one might see this as a corroboration of Hart and Honoré's thesis that choices about what the cause is aren't arbitrary. It appears that there are independent arguments for reaching a similar conclusion.

References

Beebee, H. (2011). Necessary connections and the problem of induction. *Noûs*, *45*(3), 504–527. https://doi.org/10.1111/j.1468-0068.2010.00821.x

Cartwright, N. (1983). Causal laws and effective strategies. In *How the laws of physics lie*. Oxford and New York: Clarendon Press and Oxford University Press. (Reprinted from *Noûs*, 1979, *13*; in part, new)

Cartwright, N. (1989). *Nature's capacities and their measurement*. Oxford: Clarendon Press.

Cartwright, N. (1999). *The dappled world. A study of the boundaries of science*. Cambridge: Cambridge University Press.

Cartwright, N. (2007). What is wrong with Bayes nets? In *Hunting causes and using them* (pp. 61–79). Cambridge: Cambridge University Press. (Reprinted from *The Monist*, *84*(2), 242–264 (2001))

Clatterbaugh, K. (2009). The early moderns. In H. Beebee, C. Hitchcock, & P. Menzies (Eds.), *The Oxford handbook of causation* (pp. 55–72). Oxford: Oxford University Press.

Dowe, P. (1992). Wesley Salmon's process theory of causality and the conserved quantity theory. *Philosophy of Science*, *59*(2), 195–216. https://doi.org/10.1086/289662

Dowe, P. (2000). *Physical causation*. Cambridge: Cambridge University Press.

Dowe, P. (2001). A counterfactual theory of prevention and 'causation' by omission. *Australasian Journal of Philosophy*, *79*(2), 216–226. https://doi.org/10.1080/713659223

Dupré, J. (1993). *The disorder of things. Metaphysical foundations of the disunity of science*. Cambridge, MA and London, England: Harvard University Press.

Eells, E. (1991). *Probabilistic causality*. Cambridge: Cambridge University Press.

Glymour, C., Scheines, R., Spirtes, P., & Kelly, K. (1987). *Discovering causal structure. Artificial intelligence, philosophy of science, and statistical modeling*. Orlando: Academic Press.

Good, I. J. (1961a). A causal calculus (i). *British Journal for the Philosophy of Science*, *XI*(44), 305–318. https://doi.org/10.1093/bjps/XI.44.305

Good, I. J. (1961b). A causal calculus (ii). *British Journal for the Philosophy of Science*, *XII*(45), 43–51. https://doi.org/10.1093/bjps/XII.45.43

Good, I. J. (1962). Errata and corrigenda. *British Journal for the Philosophy of Science*, *XIII*(49), 88. https://doi.org/10.1093/bjps/XIII.49.88

Hart, H. A. L., & Honoré, T. (1985). *Causation in the law* (2nd ed.). Oxford: Clarendon Press. (Original work published 1959)

Hempel, C. G. (1965). *Aspects of scientific explanation and other essays in the philosophy of science*. New York: Free Press.

Hesslow, G. (1976). Two notes on the probabilistic approach to causality. *Philosophy of Science*, *43*(2), 290–292. https://doi.org/10.1086/288684

Hitchcock, C. (2007). How to be a causal pluralist. In P. Machamer & G. Wolters (Eds.), *Thinking about causes* (pp. 200–221). Pittsburgh, PA: University of Pittsburgh Press.

Hume, D. (1999). *An enquiry concerning human understanding*. In T. L. Beauchamp (Ed.). Oxford, New York: Oxford University Press. (Critical text of the 1772 edition, first published 1748)

Hume, D. (2000). *A treatise of human nature*. In D. F. Norton & M. J. Norton (Eds.). Oxford, New York: Oxford University Press. (Critical texts of the first editions of *A treatise of human nature*, books 1–3, and *An abstract of… a treatise of Human Nature*.)

Hüttemann, A. (2013). A disposition-based process-theory of causation. In S. Mumford & M. Tugby (Eds.), *Metaphysics and science* (pp. 101–122). Oxford: Oxford University Press.

Kim, J. (1973). Causes and counterfactuals. *The Journal of Philosophy*, *70*(17), 570–572. https://doi.org/10.2307/2025312

Lewis, D. (1973). Causation. *The Journal of Philosophy*, *70*(17), 556–567. https://doi.org/10.2307/2025310

Lewis, D. (1986). *Philosophical papers* (Vol. 2). New York, Oxford: Oxford University Press.

Lewis, D. (2001). *Counterfactuals*. Malden, MA: Blackwell Publishers. (Original work published 1973)

Lewis, D. (2004). Causation as influence. In J. Collins, N. Hall, & L. A. Paul (Eds.), *Causation and counterfactuals* (pp. 75–106). Cambridge, Massachusetts and London, England: MIT Press. (Reprinted from *Journal of Philosophy*, *97*(4), 182–197 (2000); minor revisions have been made for consistency)

Mackie, J. L. (1980). *The cement of the universe. A study of causation*. Oxford: Clarendon Press.

Mill, J. S. (1973). *A system of logic ratiocinative and inductive. Being a connected view of the principles of evidence and the methods of scientific investigation. Books i–iii* (Paperback 2011 ed., Vol. VII; In J. M. Robson (Ed.). Abingdon: Routledge. (With an Introduction by R. F. McRae).

Otte, R. (1981). A critique of Suppes' theory of probabilistic causality. *Synthese*, *48*(2), 167–189.

Pearl, J. (2009). *Causality. Models, reasoning, and inference* (2nd ed.). Cambridge: Cambridge University Press. (Original work published 2000)

Pearl, J. (2014). Comment: understanding Simpson's paradox. *The American Statistician*, *68*(1), 8–13. https://doi.org/10.1080/00031305.2014.876829

Psillos, S. (2002). *Causation and explanation*. Durham: Acumen.

Reichenbach, H. (1930). Kausalität und Wahrscheinlichkeit. *Erkenntnis*, *1*, 158–188. https://doi.org/10.1007/BF00208615

Reichenbach, H. (1949). Philosophical foundations of probability. In *Proceedings of the [first] Berkeley symposium on mathematical statistics and probability* (pp. 1–20). Berkeley, California: University of California Press. Retrieved from https://projecteuclid.org/euclid.bsmsp/1166219195

Reichenbach, H. (1999). *The direction of time*. In M. Reichenbach (Ed.). Mineola, New York: Dover Publications. (Original work published 1956)

Reid, T. (1843). *Essays on the active powers of the human mind; an inquiry into the human mind on the principles of common sense; and an essay on quantity*. London: Thomas Tegg, Cheapside.

Richman, K. A. (2007). Debating the New Hume. In R. Read & K. A. Richman (Eds.), *The New Hume debate* (Revised ed., pp. 1–15). London and New York: Routledge.

Rosen, D. A. (1978). In defense of a probabilistic theory of causality. *Philosophy of Science*, *45*(4), 604–613. https://doi.org/10.1086/288838

Russell, B. (1948). *Human knowledge. Its scope and limits*. London: George Allen and Unwin Ltd.

Salmon, W. C. (1971). Statistical explanation. In W. C. Salmon (Ed.), *Statistical explanation & statistical relevance* (pp. 29–87). Pittsburgh, PA: University of Pittsburgh Press.

Salmon, W. C. (1978). Why ask, "why?"? An inquiry concerning scientific explanation. *Proceedings and Addresses of the American Philosophical Association*, *51*(6), 683–705. https://doi.org/10.2307/3129654

Salmon, W. C. (1980). Probabilistic causality. *Pacific Philosophical Quarterly*, *61*(1–2), 50–74. https://doi.org/10.1111/j.1468-0114.1980.tb00004.x

Salmon, W. C. (1984). *Scientific explanation and the causal structure of the world*. Princeton, NJ: Princeton University Press.

Salmon, W. C. (1998). *Causality and explanation* (Kindle ed.). New York, Oxford: Oxford University Press.

Salmon, W. C. (2002). Causation. In R. M. Gale (Ed.), *The Blackwell guide to metaphysics* (pp. 19–42). Oxford, Malden: Blackwell Publishers.

Salmon, W. C. (2018). A realistic account of causation. In M. Marsonet (Ed.), *The problem of realism* (ebook ed., pp. 106–134). London and New York: Routledge. (Original work published 2002). https://doi.org/10.4324/9781315185507

Schaffer, J. (2005). Contrastive causation. *The Philosophical Review*, *114*(3), 327–358. https://doi.org/10.1215/00318108-114-3-327

Schurz, G. (2001). What is 'normal'? An evolution-theoretic foundation for normic laws and their relation to statistical normality. *Philosophy of Science*, *68*(4), 476–497. https://doi.org/10.1086/392938

Simpson, E. H. (1951). The interpretation of interaction in contingency tables. *Journal of the Royal Statistical Society: Series B (Methodological)*, *13*(2), 238–241. https://doi.org/10.1111/j.2517-6161.1951.tb00088.x

Skyrms, B. (1980). *Causal necessity. A pragmatic investigation of the necessity of laws*. New Haven and London: Yale University Press.

Spirtes, P., Glymour, C., & Scheines, R. (2000). *Causation, prediction, and search* (2nd ed.). Cambridge, Massachusetts and London, England: MIT Press. (Original work published 1993)

Spohn, W. (2014). The epistemic account of ceteris paribus conditions. *European Journal for Philosophy of Science*, *4*, 385–408. https://doi.org/10.1007/s13194-014-0093-6

Suppes, P. (1970). *A probabilistic theory of causality*. Amsterdam: North-Holland Publishing Company.

Wagner, C. H. (1982). Simpson's paradox in real life. *The American Statistician*, *36*(1), 46–48. https://doi.org/10.1080/00031305.1982.10482778

Waters, C. K. (2007). Causes that make a difference. *The Journal of Philosophy*, *104*(11), 551–579. https://doi.org/10.5840/jphil2007104111

Woodward, J. (2003). *Making things happen. A theory of causal explanation*. New York: Oxford University Press.

Chapter 3
Recognising Pluralism and Arity Pluralism

Abstract This chapter introduces an account of causal pluralism via an extensional analysis of causal concepts. I first discuss well-known limitations of first-order logic and argue that it is nevertheless possible to employ first-order logic in an attempt at structuring claims of causal pluralism. I distinguish (plain) causal pluralism from arity pluralism. Plain causal pluralism is the claim that one cannot successfully analyse a master concept that subsumes more specific causal concepts. Monism is the claim that this is possible. Arity pluralism is a form of conceptual plurality that postulates causal concepts of different arity.

Keywords Arity pluralism · Causal monism · Causal pluralism · Conceptual plurality · Extensional analysis · Family resemblance · Intramural pluralism · Quasi-causation · Translation policy · Two concepts of cause

A useful approach to conceptual pluralism is paying extra—though not exclusive—attention to the extensions of concepts. The extension (another term would be 'denotation') is the 'thing' or the (abstract) object to which a word or proposition refers. The extension has to be distinguished from the intension (also referred to as 'sense'). The customary example used to illustrate the difference is that of the planet Venus, which can be referred to as 'morning star' and 'evening star'. If one considers the extension only, there is no difference between saying 'the morning star has risen' or saying 'the evening star has risen'. The intuition of there being a difference between these two expressions is addressed under the notion of intension. Although these expressions have the same extension, they differ with respect to their intension. Although no longer a surprise to us today, it would have been interesting to ancient or prehistoric people to be told that the morning star and the evening star are the same planet (Mates, 1972, pp. 22 ff.; Fitting, 2015; Frege, 1948, pp. 214 f.; Frege, 1986, pp. 46 f.).

Extension and intension are both components of the meaning of concepts. Yet within the enquiry conducted in this book, extensions are often the more important components of meaning. Imagine two researchers at the beginning of written history, one of them having as a research subject the morning star and the other the evening

K. Ehrenstein, *Causal Pluralism in the Life Sciences*,
History, Philosophy and Theory of the Life Sciences 25,
https://doi.org/10.1007/978-3-030-87942-6_3

star. If those researchers had a dispute over those 'stars', it is immensely important for both of them to know that they are in fact arguing about the *same* entity.

Followers of different causal theories argue about both extension and intension of causal concepts. The antirealist's idea of causation is different from the realist's. What these ideas are is articulated by the respective theories, i.e. these theories give an account of the intension of the causal concepts they postulate. Whether these intensions refer to the things they describe is another question. The anti-realist can, for instance, doubt that essences postulated by some realist theories exist. The realist, in contrast, may argue that anti-realist theory lacks explanatory power without such postulates. Yet where both theories can intersect is that they describe common phenomena even if they disagree about the metaphysical underpinning. From the antirealist perspective the concepts postulated by realist theory are partially empty, i.e. these theories may describe the phenomena accurately and at the same time be mistaken about the postulated metaphysical details. From the realist's perspective the objection is that concepts coming out of antirealist theory may describe the phenomena accurately, but fail to capture commonly shared beliefs such as, for instance, that handling plutonium is *inherently* dangerous.

I regard the metaphysical playground to be a distraction when one attempts to bring order to conceptual plurality. That is why I have omitted much of the metaphysical detail in Chap. 2. Hume's antirealism needs to be considered, because that is what he is famous for. But is it necessary to talk about the metaphysics of probabilities when discussing probabilistic theories of causation? And does one really need to speak about essences when giving an introduction to disposition-based process theory?

My opinion is that realist theories such as the disposition-based process theory are the more reasonable choice. Yet pushing that agenda won't help me with conducting the analysis I am about to undertake in this chapter. I will therefore adopt a methodological agnosticism with respect to the metaphysics of causation. My concern will be to analyse how concepts are related if one leaves aside the metaphysical dispute.

3.1 Extensions of Concepts of Cause and Effect

One way of bringing out differences between causal concepts is to think about possible translations into first-order predicate calculus, variously known as predicate calculus of the first order, first-order logic, or the formal language $\mathfrak{L}$.[1] Here are two example sentences for translation:

1. The lightning is a cause of the forest fire.
2. The postman's visit is a cause of the dog's barking.

[1] I thank Vera Hoffmann-Kolss for reading an early draft of this section and alerting me to mistakes.

Those propositions are the sort of causal judgements examined and articulated by the Lewisian counterfactual theory (Sect. 2.2). I shall assume that the causal relata in the above examples are events. These sentences can easily be translated into $\mathfrak{L}$ by choosing a predicate and individual constants. From here onwards I will think of causal concepts as predicates in $\mathfrak{L}$.

I shall carry out the translation of the example sentences in accordance with Mates (1972, Chap. 5). Translating a sentence from a natural language into $\mathfrak{L}$ requires giving an interpretation $\mathfrak{I}$ and choosing a sentence of $\mathfrak{L}$ that comes close to the sentence of the natural language and has the same truth conditions under $\mathfrak{I}$ (Mates, 1972, pp. 69 f.).

In simple cases a sentence of $\mathfrak{L}$ is an atomic formula. This is a predicate letter followed by individual symbols. 'C^2_2', for instance, is a predicate of the language $\mathfrak{L}$. The subscript is a means of distinguishing predicates. Hence, 'C^2_2' and 'C^2_{62}' do not denote the same predicate. The superscript tells one that they are both predicates of the degree 2. In a sentence of $\mathfrak{L}$ a predicate of the degree 2 has to be followed by two individual symbols. A predicate of the degree 1 is followed by one individual symbol, a predicate of the degree 3 is followed by three individual symbols, and so on. There are two sorts of individual symbols: individual constants and variables. Individual constants are lower-case letters from the beginning of the alphabet: 'a', 'b', 'c', and so on. One may introduce further constants by adding subscripts, e.g. 'a_7', 'd_{45}', 'e_{2005}'. So far variables are not yet needed, but one may use the lower-case letters from the end of the alphabet (u–z) as variables and introduce further variables by adding subscripts. Wherever a variable occurs in a sentence of $\mathfrak{L}$ it has to be bound by a quantifier. There are two quantifiers in $\mathfrak{L}$: the universal quantifier '$\forall$' and the existential quantifier '$\exists$'. In this section I am only concerned with atomic formulae (or atomic sentences). However, by using the symbols '$\sim$' (not), '$\vee$' (or), '$\wedge$' (and), '$\rightarrow$' (if), or '$\leftrightarrow$' (iff = if and only if) one may compose more complex sentences (Mates, 1972, pp. 44 ff.). I shall create more complex sentences in Chap. 4.

In order to give an interpretation one has to specify the universe of discourse, i.e. the set $\mathfrak{D}$. The choice of the universe of discourse is fairly arbitrary: there are virtually no restrictions except that $\mathfrak{D}$ mustn't be the empty set. The set $\mathfrak{D}$ contains anything one wants to talk about, i.e. those objects (this includes abstract entities) of which one actually talks and those objects of which one could talk if one wanted to. More technically speaking, the elements of $\mathfrak{D}$ are those objects to which one can refer by means of individual constants or variables (Mates, 1972, pp. 55 f.).

Once the universe of discourse is specified, denotations are assigned to the non-logical constants (Mates, 1972, p. 56). The non-logical constants are the predicate letters and the individual constants (Mates, 1972, p. 45). To each individual constant one assigns an element of $\mathfrak{D}$. To each predicate one assigns sets of elements of $\mathfrak{D}$ or sets of n-ary relations between elements of $\mathfrak{D}$. To a predicate of degree 1 one assigns a set of elements of $\mathfrak{D}$, to a predicate of degree 2 one assigns a set of binary relations between elements of $\mathfrak{D}$, to a predicate of degree 3 one assigns a set of ternary relations between elements of $\mathfrak{D}$, and so on (Mates, 1972, pp. 55 f.). The variables in first-order logic allow one to speak about elements of $\mathfrak{D}$ in general

(Mates, 1972, pp. 44 ff.). These variables must be distinguished from other variables below that I shall call 'Woodward-variables'.

Once all these decisions are made, predicates and individual constants may be put together in order to form sentences in $\mathfrak{L}$. The next step is assigning truth-values to the sentences. In order to assign the truth value one has to settle the question of whether the n-ary relation of which one speaks in a sentence in $\mathfrak{L}$ is an element of the set that has been assigned to the predicate. One has to ask, for instance, whether $< a, b >$ is an element of the set that is assigned to the predicate C_2^2. If it is, one assigns the truth-value 'true'; if it is not, one assigns the truth-value 'false' (Mates, 1972, pp. 55 f.).

In order to translate the two example sentences provided at the beginning of this section, I give the following interpretation $\mathfrak{I}$, i.e. I stipulate extensions as follows:

$\mathfrak{D}$: the set of token events
C_2^2: ① is a token cause of ②[2]
a: the event that the lightning occurs
b: the event that the forest fire occurs
c: the event that the postman visits
d: the event that the dog barks

Having stipulated this, I am now in a position to represent the two example sentences formally as

$$C_2^2 ab \tag{3.1}$$

and

$$C_2^2 cd\ . \tag{3.2}$$

These sentences in $\mathfrak{L}$ are translations of 'the lightning is a cause of the forest fire' and 'the postman's visiting is a cause of the dog's barking'. If one assumes that the set I assigned to C_2^2 contains $< a, b >$ and $< c, d >$ as elements (among other elements that are also binary relations), both sentences turn out to be true relative to the interpretation $\mathfrak{I}$.

I hope the reader will agree that—so far—nothing interesting has happened. My translation of the above sentences proceeded smoothly. But now, let me turn to another example. Consider the following causal statement (in terms of 'Woodward-variables'):

X is a direct cause of Y with respect to the variable set $\mathbf{V}$.

---

[2] This is a simplified notation used in Mates (1972, Chap. 5). Writing 'C_2^2: ① is a token cause of ②' means that I assign the set of binary relations {< ①, ② > | ① is a token cause of ②} to C_2^2. The counter '①' always refers to the first element of an n-tuple, the counter '②' always refers to the second element of an n-tuple, the counter '③' always refers to the third element of an n-tuple, and so on.

How can I translate this into $\mathfrak{L}$? (The definition of a direct cause was quoted in Sect. 2.4.) A fundamental question concerns the universe of discourse. According to the account given in Sect. 2.4, the causal relata in an interventionist theory are thought to be variables. Woodward explicates the term this way: 'variables are properties or magnitudes that [...] are capable of taking more than one value' (Woodward, 2003, p. 39). I call these variables 'Woodward-variables' in order to distinguish them from variables as they are used in first-order logic. It would appear that the universe of discourse has to contain Woodward-variables at least. Yet those variables are not the only entities mentioned in the above statement: there is also the variable set **V**.

At this point there are two ways that one could proceed with the translation: one may suppose that **V** is an element of the universe of discourse or one may suppose that **V** itself *is* the universe of discourse. The latter suggestion requires *rethinking* what is written as a causal statement above. One could suggest that the causal statement is 'X is a direct cause of Y' rather than 'X is a direct cause of Y with respect to the variable set **V**'. If one took that view, one could argue that the variable set **V** (which is obviously a set of Woodward-variables) is *identical* to the universe of discourse. From a formal point of view I would concede that this is a possible way of specifying the universe of discourse. The suggestion would be to give the interpretation $\mathfrak{I}'$

$\mathfrak{D}'$ $= \mathbf{V}$
C^2_{62}: ① is a direct cause of ②
a_2: the Woodward-variable X
b_2: the Woodward-variable Y

and to translate 'X is a direct cause of Y' into $\mathfrak{L}$ with the sentence

$$C^2_{62}a_2b_2 \,. \tag{3.3}$$

But although this is possible from a *purely* formal perspective, I disagree with it from a translator's perspective. The point of a translation is not to produce just *anything* that is in accordance with formal rules, but rather to produce a translation that is *close to the original and consistent with a general policy of making translations*. I think the decision to choose **V** as the universe of discourse violates both principles: firstly, it disregards the apparent structure of the above statement, which relates *three* entities and not just two; secondly, it is also *inconsistent* with the above-established policy of specifying the universe of discourse.

The latter point may not be obvious. In the above example the universe of discourse $\mathfrak{D}$ is the set of token events. Stated like this, the specification of the set is a bit vague. However, it is quite easy to give a more precise account: if one thinks of it in the context of the Lewisian counterfactual analysis, it appears that $\mathfrak{D}$ should contain all past, present and future events of the actual world and of any possible world. In other words, the universe of discourse is specified in a way that includes *all* candidate events of which one might want to say that they are a cause of another event or the effect of another event. Note that the interpretation $\mathfrak{I}'$ does *not* accord with this policy. $\mathfrak{I}'$ specifies $\mathfrak{D}'$ more narrowly: $\mathfrak{D}'$ is a set that includes the Woodward-variable X, the Woodward-variable Y, and those Woodward-variables of which it is said in the definition of a direct cause that they are to be 'held fixed at some value by interventions' (Woodward, 2003, p. 55). This is as if one had specified $\mathfrak{D}$ as the set that contains a, b, c, and d only. $\mathfrak{D}'$ is by comparison much more restricted than $\mathfrak{D}$.[3] Identifying **V** with $\mathfrak{D}'$ is an inconsistent translation policy.

It is now obvious that the universe of discourse has to be specified differently. The universe of discourse has to be more inclusive. It should be something like the set containing all Woodward-variables one can think of and all sets of Woodward-variables that are of the **V**-sort (with **V** being just one among many other similar sets). This set is a more heterogeneous set than $\mathfrak{D}$, but it exhibits *a similar degree of generality*.

After specifying the universe of discourse this way it is natural to suggest further changes: the most significant is to represent the concept of a direct cause as a ternary relation instead of a binary relation. Hence, I give the revised interpretation $\mathfrak{I}''$:

$\mathfrak{D}''$: the set of Woodward-variables and **V**-like sets of Woodward-variables
C^3_{64}: ① is a direct cause of ② with respect to the variable set ③
a_4: the Woodward-variable X
b_4: the Woodward-variable Y
c_4: the set **V**

This interpretation allows one to translate 'X is a direct cause of Y with respect to the variable set **V**' into $\mathfrak{L}$ with the sentence

$$C^3_{64}a_4b_4c_4\ . \tag{3.4}$$

This translation is in my view preferable to the translation that makes use of 'C^2_{62}' instead of 'C^3_{64}'. As a solution, it has two advantages: firstly, it reproduces the conspicuous structure of the statement more accurately, i.e. it represents it as a ternary relation instead of a binary relation; and secondly, the interpretation $\mathfrak{I}''$ stipulates a universe of discourse which avoids the inconsistency identified in the previous translation policy.

What I have discussed here belongs to the basic teachings found in introductions to first-order logic. In the textbook by Paul Hoyningen-Huene (1998, p. 172) one finds a discussion of an analogous case. The sentence he uses as an example is 'Lake Constance lies between the North Sea and the Mediterranean Sea'. How would one represent its predicate structure? In an attempt to translate the sentence one may choose between several options. Here is a list of some predicates one could think of

[3] Note that this comparison abstracts from the fact that 'a', 'b', 'c' and 'd' refer to events and 'X' and 'Y' refer to changes in properties or magnitudes (or, in the case of indicator variables, also events).

while translating the above sentence into $\mathfrak{L}$:

L_7^1: ① lies between the North Sea and the Mediterranean Sea
L_8^2: ① lies between the North Sea and ②[4]
L_9^3: ① lies between ② and ③

Returning to our two original example sentences, it is not even true that the translations of 'the lightning is a cause of the forest fire' and 'the postman's visit is a cause of the dog's barking' are *un*ambiguously adequate. I could have suggested using the following predicates instead of 'C_2^2':

F^1: ① is a cause of the forest fire
D^1: ① is a cause of the dog's barking

These predicates also allow a translation into $\mathfrak{L}$. Suppose $\mathfrak{D}$ is still the universe of discourse and the individual constants introduced above are the same. Then I can write

$$F^1 a \tag{3.5}$$

and

$$D^1 c\,. \tag{3.6}$$

There are, however, two deeply disturbing ways in which these translations appear to be inadequate: firstly, it seems that both cause *and* effect should be recognisable as entities in their own right; secondly, one would expect the use of the same predicate in both sentences. These translations, however, seem to presuppose that there are *no effects* and that there are *distinct concepts of cause*.

But that is not how we think of the concept of cause. When I say that the lightning is a cause of the fire and that the postman's visit is a cause of the dog's barking, I am supposing that the relation between these events is of the *same* sort. It is for this reason that I stipulated in the above interpretation $\mathfrak{I}$ that both $< a, b >$ and $< c, d >$ are elements of the *same* set (that I assigned to C_2^2). The translations shown here disregard that belief. They are not wrong, strictly speaking, but they head in the wrong direction. One doesn't usually want to say (although doing so wouldn't be wrong) that the lightning has the property of causing the fire or that the postman's visit has the property of causing the dog's barking. In other words, both translations are inadequate.

As can be seen here, unambiguously identifying a predicate structure presents a general problem. One faces the difficulty that one can choose more or less arbitrarily between translations into $\mathfrak{L}$. Nevertheless, I think that my discussion in

[4] A further option would be to switch the North Sea with the Mediterranean Sea in this sort of predicate.

the preceding paragraphs has also shown how to argue that some translations into $\mathfrak{L}$ are more adequate than other translations, even though it isn't a matter of being strictly right or wrong.

A conclusion to be drawn from this section is that one needs a consistent translation policy if one wants to speak about the arity of concepts of cause and effect, or if one wants to compare concepts of cause and effect with respect to their arity (otherwise it will be relatively uninformative to speak about the arity of a concept at all). That translation policy may be described as follows: One has to choose the universe of discourse in a manner that includes all candidate effects and candidate causes. One also has to include entities such as the set **V** and its peers (**V**-like sets). In certain cases, **V**-like sets could be background circumstances, causal fields and so on. However, this alone is not enough. Predicates also need to be chosen in a way that makes use of these modifications. Specifying the universe of discourse more inclusively without modifying the predicates accordingly would be pointless.

Provided one adopts this policy and sticks to it, one is in a position to make tentative suggestions about the arity of concepts of cause and effect. In the following discussion I shall make such suggestions, although I must emphasise that they are not and *cannot be* ironclad claims. The outcome of the argument in this section is that *the extensions of concepts of cause and effect are not unambiguously determinable*.

3.2 Recognising Pluralist Accounts

Armed with first order logic, I shall now try to distinguish monist from pluralist positions. My suggestion is to think of monist theories as approaches that analyse a causal master concept and subsume all other causal concepts to this general concept. Pluralist theories, by contrast, analyse one or more causal concepts and subsume them under a general causal concept but without analysing that general concept. By 'subsumption' I mean that the extension of the subsumed concept is a subset of the extension of the subsuming concept.

Note that the criterion for distinguishing monism from pluralism is *not* how many concepts there are. That would trivialise the notion of pluralism, since there are undoubtably many causal concepts. I ask instead where the target of the analysis is positioned within a concept hierarchy. According to this criterion, a pluralist position need not analyse more than one concept. If the only target of the analysis ranks low in the concept hierarchy, this is pluralism. A monist theory, on the other hand, can in principle analyse more than one concept as long as this includes the master concept on top of the hierarchy.

A good example of monism is Lewis's counterfactual analysis (Sect. 2.2). According to that position, the concept of causation is 'applicable to all different kinds of causation, and applicable even to kinds of causation never found in our own world' (Lewis, 2004, p. 76). Lewis's analysis takes into account various counterfactual scenarios, some of them at a considerable distance from actuality, e.g. scenarios that involve magic (Lewis, 2004, p. 81). He argues that his conceptual, i.e. counterfactual, analysis should extend to other modes of causation, because we are able to *understand* fictional stories 'in which causation works by magical mechanisms entirely alien to the world of our acquaintance' (Lewis, 2004, p. 76). According to Lewis, 'conceptual analysis is required to reveal what it is that all the actual and possible varieties of causation have in common' (Lewis, 2004, p. 76). He says that '[i]f causation is, or might be, wildly disjunctive, we need to know what unifies the disjunction' (Lewis, 2004, p. 76). Lewis's counterfactual analysis is supposed to unify this disjunction. It defines (or perhaps only explicates) when a binary relation between events is a causal relation. I see this as a paradigm example of monism.

Next, let us look at an example of pluralism. It has been argued in recent years that counterfactual analysis fails. Ned Hall (2004), the major proponent of this view, thinks that one cannot carry out a counterfactual analysis. He argues that there is an insurmountable tension between what he sees as the five defining features of causation: the three desiderata, i.e. transitivity, locality and the intrinsicness thesis, that define the concept of *production*; and the two desiderata, i.e. dependence and the omissions thesis, that define the concept of *dependence*.

Hall aims to show that there are counterexamples to counterfactual analysis which reveal that these two concepts *must* come apart under certain circumstances. One of those examples (Hall, 2004, pp. 241 f.) involves a bomber being escorted by a fighter jet, and an enemy aircraft intercepting the bomber and fighter jet. Hall argues that it is impossible to postulate any local connection in the sense that locations of the fighter jet in the actual world intersect with space occupied by the enemy aircraft in the nearby possible worlds. The jets anticipate each other's movements in such a way that they are always where the other aircraft is not in the nearby possible worlds (Fig. 3.1).

Hall concludes from this and other test cases that any analysis of the overarching concept of causality 'is doomed to failure' (Hall, 2004, p. 256). Note that Hall still acknowledges that there is an overarching concept in the sense that ordinary language doesn't differentiate between production and dependence. He explains this by arguing that 'it should come as no surprise that the distinction between production and dependence has gone unnoticed, for *typically* the two relations coincide' (Hall, 2004, p. 254, italics in original). This means that the extensions of 'production' and 'dependence' intersect. In many cases both concepts are applicable, i.e. they yield the same truth value. In other cases, only one or the other can be used. If one applies both concepts in such a situation anyway, the truth values come apart.

Hall's position is a paradigm example of pluralism, since he argues that the master concept cannot be analysed and that an analysis must therefore aim at the two intersecting subsets of the master concept.

Although Lewis's position is easily categorised as monism and Hall's as pluralism, it can be less obvious with which side a position should be grouped. The conserved quantity theory by Dowe is an example (Sect. 2.5).

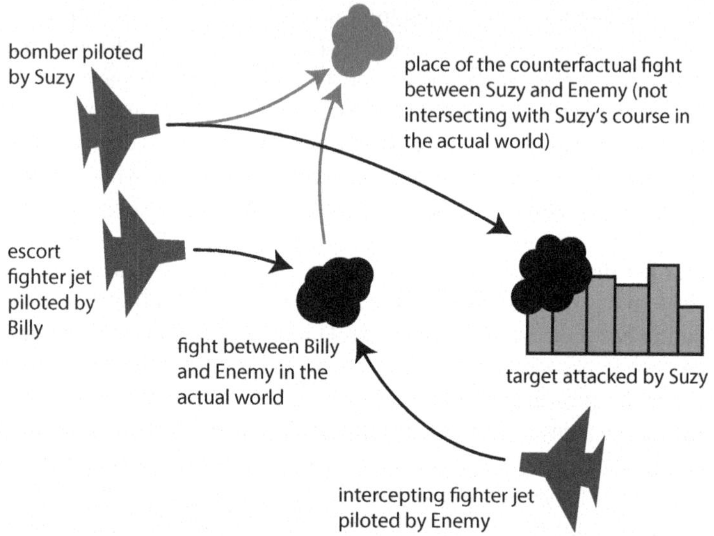

Fig. 3.1 My illustration of Hall's counterexample in which locality fails. In the actual world Suzy is accompanied by Billy who shoots down an enemy fighter jet, which intercepts Suzy in possible worlds in which Billy loses the fight against Enemy. Note that in these possible worlds Suzy's flight path is different since she tries to evade Enemy. The place of the counterfactual fight between Suzy and Enemy doesn't intersect with Suzy's actual flight path

Dowe puts forward the slogan 'horses for courses' (Dowe, 2000, p. 12). He says that 'we cannot assume that the best conceptual analysis is also the best empirical analysis, or vice versa' (Dowe, 2000, p. 12). He describes the difference between a conceptual analysis and an empirical analysis as a difference between an *a priori* analysis and an *a posteriori* analysis. According to Dowe, the conceptual analysis 'is concerned to spell out the logical consequences and to propose a plausible and illuminating explication of the concept' (Dowe, 2000, p. 2), while the '[e]mpirical analysis aims to map the objective world, not our concepts' (Dowe, 2000, p. 3). The former analysis, if it succeeds, yields an *a priori* truth; the latter analysis, if it succeeds, yields an *a posteriori* truth (Dowe, 2000, pp. 2 f.).

Nevertheless, both analyses explicate a concept of cause: there is the concept of cause, as explicated by the conceptual analysis, and there is the concept of cause, as explicated by the empirical analysis. Both concepts have an extension—no matter how one arrives at them, an extension is an extension, regardless of the analysis (*a priori* or *a posteriori*) that one uses in order to delineate the extension. What is the relation between these extensions? Is one a subset of the other? In other words, does one concept subsume the other?

A correct, although unhelpful, answer would be to point out that the two extensions do not intersect for the *trivial* reason that one concept (Lewis's concept of cause) is a relation between events, while the other concept (Dowe's concept of cause) is a relation between processes. So let us assume, for the sake of argument, that there is no difference between events and processes, and no need to say that the two extensions are trivially non-intersecting. If one makes that auxiliary assumption, it seems that Lewis's concept of cause subsumes Dowe's concept of cause: Lewis's analysis extends to all possible worlds, while Dowe's analysis is confined to the actual world. If this were true, Dowe's account would have to be classified as a pluralist position for the reason that the examined concept is not at the top of the hierarchy. I believe that Dowe's slogan 'horses for courses' invites this reading.

Yet this isn't the only possible interpretation. There is also the fact that Dowe distinguishes 'causation' from 'quasi-causation' (Sect. 2.5). This terminology strongly invites reading Dowe as if he believed that Lewis hasn't got a concept of cause in a strict sense. If it is part of Dowe's theory to deny the legitimacy of Lewis's analysis, then no subsumption to a broader concept occurs. If this were true, it would be natural to suggest that Dowe's concept of cause occupies the top-position instead.

One would then have to group Dowe with monist positions. In order to cope with this ambiguity the position could be classified as quasi-pluralism[5] (which isn't pluralism in a strict sense).

Another position that is easily categorised is Cartwright's view of causation. She thinks that causation is not 'a single monolithic concept' (Cartwright, 2004, p. 813). This view gained popularity through G. E. M. Anscombe who argued that '[t]he word "cause" itself is highly general' (Anscombe, 1993, p. 93). One could *add* such a general word to a language that already has various special concepts of cause. But a language that hasn't got special concepts of cause wouldn't have a general concept of cause (Anscombe, 1993, p. 93).

The Anscombeian view is influential in the literature on biological mechanisms. It is a central supposition in Peter Machamer, Lindley Darden, and Carl Craver's account of a mechanism (2000, p. 6). The Anscombeian view of causation is defended at length by Machamer (2004). He thinks that

> [t]he problem of causes is not to find a general and adequate ontological or stipulative definition, but a problem of finding out, in any given case, what are the possible, plausible, and actual causes at work in any given mechanism (Machamer, 2004, pp. 27 f.).

This view exemplifies pluralism at its highest degree: Machamer suggests focusing at a basic level. According to Machamer et al. (2000), one has to analyse mechanisms in terms of both entities (which are thought to be things like ions or membranes) and activities (which are thought to be causes like repelling or bonding). One of their major arguments is that this was 'descriptively adequate' (Machamer et al. 2000, p. 23). One could say that their thesis is that analyses of concepts of cause mustn't be too abstract.

Cartwright distinguishes between 'thick' concepts of cause and 'thin' concepts of cause (Cartwright, 2004, pp. 814 ff.). The thick concepts are those concepts that specify or describe a cause in a very concrete way. They are used in statements such as 'the carburetor *feeds* gasoline and air to a car's engine' or 'the low-pressure air

[5] This expression was suggested to me by Andreas Hüttemann in analogy to Dowe's notion of quasi-causation.

sucks gasoline out of a nozzle' (Cartwright, 2004, p. 815, italics in original). The thin concepts of cause are those concepts that are subjects of typical philosophical analyses of causation. Cartwright thinks that these thin concepts are important for scientific practice. They are particularly useful when one wants to make causal inferences. But in her view it wouldn't be right to claim that thick causal concepts are replaceable with thin causal concepts (Cartwright, 2004, pp. 817 f.).

There is a tacit background assumption in the sources discussed here which is never made fully explicit, and that is that relations between causal concepts are to be thought of in terms of family resemblance. This was a notion made popular by Ludwig Wittgenstein. His paradigm example for illustrating family resemblance is the concept of a game. Wittgenstein considers the similarities between board games, card games, ball games and other sorts of game. This lets him conclude that there is no commonality between all sorts of game: some games are entertaining, others are not; some games are competitive, others are not; some games are played in company, others are not; and so on. As a consequence, Wittgenstein maintains that one shouldn't say that there is a commonality between all sorts of game, but that those things that are called 'game' are related to each other in 'a complex net of resemblances' (ein kompliziertes Netz von Ähnlichkeiten). He speaks of those similarities as 'family resemblances' (Familienähnlichkeiten), and suggests that the resemblances between games are like the resemblances between the members of a biological family, in which members resemble each other with respect to various traits such as eye colour, hair colour, stature, temperament, and so on. But—and this is the central point—there is nothing that *all* family members share with each other: A and B, for instance, may resemble each other (but not C) with respect to eye colour; B and C may resemble each other (but not A) with respect to hair colour; and A and C may resemble each other (but not B) with respect to stature, and so on (Wittgenstein, Philosophische Untersuchungen, §§ 66+67).

Another metaphor Wittgenstein uses in this context is the spinning of yarn. Wittgenstein argues that the strength of the yarn is not owed to a single fibre running over the whole length of the yarn but that the strength of the yarn is owed to the fact that many short fibres are entwined with each other. The fibres in this sort of metaphor are compared to the similarities and resemblances that hold between the members of families (Philosophische Untersuchungen, § 67).

Wittgenstein maintains that it is mistaken to postulate a property that is common to all members of such families. He considers that one could postulate a *disjunctive* family property. But he thinks that it would say very little, since it would amount to the thesis that there is a something that runs over the whole length of the yarn, which would be 'the seamless entwinement of these fibres' (das lückenlose Übergreifen dieser Fasern). Wittgenstein dismisses this option as a play on words (Philosophische Untersuchungen, § 67).

What is sometimes confusing is how philosophers in the tradition of Wittgenstein can fall back into thinking that there are overarching similarities. Anscombe suggests that there is something *common* to cause-effect relations. Her view is that 'causality consists in the derivativeness of an effect from its causes' (Anscombe, 1993, pp. 91 f.). She thinks that '[t]his is the core, the common feature, of causality in its various kinds' (Anscombe, 1993, p. 92). The paradigm example of this kind of causal relation is 'physical parenthood' (Anscombe, 1993, p. 92). In that case 'the derivation is material, by fission' (Anscombe, 1993, p. 92). It is hard to make sense of this commitment to derivativeness. Perhaps one should simply conclude that it isn't easy to hold a thoroughgoing family-resemblance view of concepts of cause and that even Wittgensteinian philosophers are tempted to search for overarching similarities. (The fact that the idea of family resemblance is only implicit in Anscombe's discussion doesn't contribute to a clarification of that question.)

In other cases it is easier to see what authors mean when they invoke the idea of family resemblance. Hart and Honoré embrace the doctrine of family resemblance, too. They think 'that there is not a single concept of causation but a group or family of concepts' (Hart & Honoré, 1985, p. 28). In their opinion these concepts 'are united not by a set of common features but by points of resemblance, some of them tenuous' (Hart & Honoré, 1985, p. 28). But although Hart and Honoré adopt the family-resemblance thesis, they believe that it is still worthwhile to give an account of a central notion of cause. Their normality account of causation does not describe what is common to *all* members of the family of causal concepts but it still describes *a commonality shared by a large number* of family members. Yet this position is still pluralism, since their theory does not suppose that this central notion of cause subsumes all other concepts.

To sum up, I have suggested that the monist need *not* deny that there is a plurality of concepts of cause. I think it is trivially true that there is a plurality of concepts of cause. It is more interesting to ask how the concept articulated by a theory fits into a hierarchy of concepts. Are the analysed concepts subsumed by more general ones, or is the analysed concept the subsuming concept?

3.3 Recognising Arity Pluralist Accounts

I shall now discuss a position that has been called 'intramural pluralism' (Hitchcock, 2007, p. 207). The central idea, as I understand it, is admitting a conceptual plurality without transcending certain boundaries. Another metaphor (or idiomatic expression) used by Hitchcock is to speak of certain concepts as if they were 'cut from the same cloth' (Hitchcock, 2001, p. 363). It appears that one may interpret 'same cloth' as 'same metaphysical nature' or 'same metaphysical kind'. Hitchcock argues that intramural pluralism 'is consistent with a monism about the underlying nature of causation' (Hitchcock, 2007, p. 210).

I shall argue that 'intramural pluralism' isn't pluralism according to the account I've developed above. 'Intramural pluralism' turns out to be a monist position (which shouldn't come as a surprise, since Hitchcock is sceptical of any pluralist thesis concerning the metaphysical nature of causation). Yet I think it is also too simple to say that 'intramural pluralism' is just a species of monism. It is different from monism, but it does not accord with my account of pluralism. My suggestion is to think of 'intramural pluralism' (or rather parts of it) as arity pluralism.[6]

Arity pluralism is a position postulating two or more causal concepts that differ with respect to arity (if thought of in terms of first-order predicate calculus). Arity pluralism and (plain) pluralism are not mutually exclusive. Mixed positions are conceivable.

However, before I explain why Hitchcock's position falls into the category of arity pluralism I shall first develop an argument against the idea that interventionism is (plain) pluralism.

Hitchcock's paradigm example of so-called 'intramural pluralism' is Eells's (1991) type-level theory of probabilistic causation. Note that I have to give out the usual warning: this is an interpretation. One doesn't find the word 'pluralism' or the words 'intramural pluralism' in Eells (1991). However, it is obvious that one finds conceptual distinctions between being a positive cause, being a negative cause, being causally neutral, being a mixed cause, and being causally relevant. Hitchcock quotes Eells in the following form:

> C is a *positive cause* (or *cause*) of E if and only if $P(E|C\&B) > P(E|{\sim}C\&B)$ for every background context B.
> C is a *negative cause* of E (or C *prevents* E or C *inhibits* E) if and only if $P(E|C\&B) < P(E|{\sim}C\&B)$ for every background context B.
> C is *causally neutral* for E (or *causally irrelevant* for E) if and only if $P(E|C\&B) = P(E|{\sim}C\&B)$ for every background context B.
> C is a *mixed cause* (or *interacting cause*) of E if it is none of the above.
> C is *causally relevant* for E if and only if it is a positive, negative, or mixed cause of E; i.e., if and only if it is not causally neutral for E. (Eells, 1991, p. 144; as paraphrased by Hitchcock, 2012, Sect. 2.6, italics in original; see also Hitchcock, 2007, p. 207)

In order to discuss whether those conceptual distinctions are an instance of (plain) pluralism I suggest translating them into the following predicates in first-order logic:

C^3_{30}: ① is a positive cause of ② with respect to revised Eells-context ③
C^3_{32}: ① is a negative cause of ② with respect to revised Eells-context ③
C^3_{34}: ① is causally neutral for ② with respect to revised Eells-context ③

As you can see, I have decided to assign sets of ternary relations to those predicates. This is a pragmatic decision. I have already discussed the problem of determining the arity of concepts of cause. Those decisions are to a certain extent

[6] I am grateful to an anonymous reviewer for suggesting this name for the position. In the original manuscript I called this 'a species of causal relativism', which was a tortuous attempt to indicate its difference from the account of plain pluralism suggested in Sect. 3.2. 'Arity pluralism' is, however, the much better name, since it avoids all the difficulties that come with the name 'relativism'—such as the connotation that there are equally valid viewpoints (cf. Boghossian (2006) for an analysis of the core theses of relativism). In the original manuscript I constantly had to remind the reader of the technical meaning I had assigned to the name in place of the meaning generally associated with the word. This is now no longer a problem and I wonder how I could have missed this solution before.

negotiable: they can be adequate, or inadequate, but not—strictly speaking—right or wrong.

Note that this time the problem isn't the decision of whether it is more appropriate to assign sets of binary relations or sets of ternary relations. According to Eells, 'type-level probabilistic causation is a relation among *four* things: a cause factor, an effect factor, a token population within which the first is some kind of cause of the second, and, finally, a kind (of population) that is associated with the given token population' (Eells, 1991, p. 22, italics added). The above reconstruction is open to the criticism that I should have chosen predicates of the degree 4, but I believe it was out of the question for me to have chosen predicates of the degree 2. In any case, my following argument will apply, no matter whether I assign ternary or quaternary relations. So I choose ternary relations to avoid distraction.

Nevertheless, keep in mind that what I call 'revised Eells-context' is an internally complex entity, which mustn't be confused with comparatively simple entities such as **V**-like sets (which are collections of Woodward-variables). The expression 'revised Eells-context' in the above reconstruction derives from the name Eells applies to his account: 'revised context approach' (Eells, 1991, p. 131). (In due course, I shall return to Eells's motivation for proposing a revised approach.) Following the example of my earlier discussion of the concept of a direct cause (Sect. 3.1) I shall refer to the causal relata in this approach by the term 'Eells-variables'. Just as with Woodward-variables one must not confuse Eells-variables with first-order logic variables.

Let me now discuss Eells's concepts of cause as I have reconstructed them here. You will have noticed that the list is not yet complete. There are two further concepts: being a mixed cause and being causally relevant. So why haven't I included them? One reason to pause at this point is to highlight the rather *casual* introduction of the concepts that follow. The definitions[7] of a mixed cause and of being causally relevant are given exclusively in terms of the preceding three concepts (positive cause, negative cause, causally neutral).

Hence, the following question arises for translations into $\mathfrak{L}$: should I eliminate the concept of a mixed cause or should I introduce a fourth and fifth predicate? Suppose that I wanted to translate a causal judgement such as 'X is a mixed cause of Y with respect to revised Eells-context R'. If I want to express this in $\mathfrak{L}$, there is—strictly speaking—no need to introduce a fourth predicate.

Assume that the universe of discourse is

$\mathfrak{D}^E$: the set of Eells-variables and revised Eells-contexts

[7] Technically speaking, the explanation of a mixed cause is no definition. It states a sufficient condition only. Yet it seems that one has to treat it as a definition. The use of 'if' in Hitchcock (2012, Sect. 2.6) is faithful to Eells (1991, p. 144) but in Hitchcock (2007, p. 207) this is written as 'iff', which seems to be more appropriate.

and that there are the individual constants:

a_6: the Eells-variable X
b_6: the Eells-variable Y
c_6: the revised Eells-context R

Then I could express the causal judgement that the Eells-variable X is a mixed cause of the Eells-variable Y given the revised Eells-context R with the following sentence in $\mathfrak{L}$:

$$\sim (C^3_{30}a_6b_6c_6 \vee C^3_{32}a_6b_6c_6 \vee C^3_{34}a_6b_6c_6)\ . \tag{3.7}$$

This is a translation of 'it is not true that X is a positive cause of Y given R or that X is a negative cause of Y given R or that X is causally neutral given R'. Of course, this is not the only option. If you dislike the above translation, you may introduce

C^3_{36}: ① is a mixed cause of ② with respect to revised Eells-context ③

as a fourth predicate. This allows you to propose the alternative translation

$$C^3_{36}a_6b_6c_6\ . \tag{3.8}$$

The obvious advantages of this latter translation are that it avoids the comparatively complex syntax of the former translation and that it takes the idea seriously that there are *four* concepts of cause: being a positive cause, being a negative cause, being causally neutral, and being a mixed cause.

But although this latter translation may be preferable for these reasons, a further difficulty is revealed. In Sect. 3.1 I argued that extensions of concepts of cause and effect are not unambiguously determinable. Now that argument has been taken to the next level: there is not just the difficulty of determining the extension of concepts but also the difficulty of determining how many concepts there are in the first place.[8]

I can reframe the above sort of argument concerning the concept of a mixed cause as an argument concerning the concept of being causally relevant. According to the definition, saying that 'X is causally relevant with respect to revised Eells-context R' is the same as saying that 'X is *not* causally neutral for Y with respect to revised Eells-context R'. Hence, I can write

$$\sim C^3_{34}a_6b_6c_6 \tag{3.9}$$

[8] That problem was partly anticipated in Sect. 3.1 where I considered replacing 'C^2_2' with 'F^1' and 'D^1'. However, in the above discussion I was simultaneously (and primarily) concerned with the arity of concepts. Now I am in a position to point out that the problem arises independently of the question of arity. In the above discussion I replaced a predicate of the degree 2 with two predicates of the degree 1. Here a predicate of the degree 3 is replaced with other predicates of the same degree.

or I can introduce the predicate

C^3_{38}: ① is causally relevant for ② with respect to revised Eells-context ③

and write

$$C^3_{38}a_6b_6c_6\ . \tag{3.10}$$

This example raises the question of how to translate into $\mathfrak{L}$ more forcefully: to suggest writing '$C^3_{38}a_6b_6c_6$' instead of '$\sim C^3_{34}a_6b_6c_6$' is almost trivial. I believe that this discussion of 'intramural pluralism' shows one how right it was to discard the idea that the pluralist postulates a plurality of concepts. A plurality of concepts is easy to come by. However, the question remains: is 'intramural pluralism' pluralism? According to my account in Sect. 3.2, monist theories analyse the master concept at the top of the hierarchy, while pluralist theories don't claim to analyse the concept at the top of the hierarchy. So, how are the above concepts related to each other?

The relations are as follows: C^3_{38} (causally relevant) subsumes C^3_{30} (positive cause), C^3_{32} (negative cause), and C^3_{36} (mixed cause). The concepts C^3_{34} (causally neutral) and C^3_{38} (causally relevant) are defined as contradictory opposites, so that neither subsumes the other. The concept that stands out is C^3_{38} (causally relevant). It sits at top of the hierarchy if one ignores C^3_{34} (causally neutral).

According this analysis, Eells's position could be monism. There is, however, one argument against this. Eells conducts *two* analyses:

> Conditional probability comparisons are for the analysis of type-level probabilistic causal relations, and not for the analysis of token-level probabilistic causal relations; and probability trajectories are for the analysis of token-level probabilistic causal relations, and not for the analysis of type-level probabilistic causal relations. (Eells, 1991, p. 5)

Eells's position is that type- and token-level concepts required different analyses, a thesis viewed critically by Hitchcock who complains that these 'theoretical accounts are so different that it is hard to see them as part of a single probabilistic taxonomy of causation' (Hitchcock, 2007, p. 212). Consequently, Hitchcock decides that Eells's position must be 'extramural pluralism', which is a term he reserves for theories that are in his opinion less acceptable, such as Hall's two-concept view.

Hitchcock also points out that he is unsatisfied with Eells's solution of offering two separate accounts. He says he is concerned

> that the distinction between type and token causation actually conflates a number of other distinctions (including the distinction between net and component effects), and that once these distinctions are disentangled, *one is left with a version of intramural pluralism* of the sort discussed above. (Hitchcock, 2007, pp. 212 f., italics added)

The difference is then that Eells doesn't analyse any master concept, while Hitchcock thinks that one can somehow subsume one level to the other, i.e. the token level to the type level. Hence, Eells's position is pluralistic and Hitchcock's position is monistic.

I think there is a further argument to show that categorising Hitchcock's position as monism is intuitively the right conclusion (and not just a consequence of my specific account of pluralism).

Hitchcock draws attention to the fact that there are colloquial counterparts to the concepts of the Eells-terminology, e.g. 'prevents' is a colloquial substitute for 'is a negative cause of' (Hitchcock, 2007, p. 207). He is, however, only concerned with the type-level concepts. But in my opinion it is more telling—or more *revealing*—to consider token-level colloquial counterparts. Consider the following three statements:

1. Fred Flintstone caused the fire.
2. Wilma Flintstone prevented the disaster.
3. Betty Rubble didn't affect the happenings in the Flintstone's house.

It appears that the first sentence corresponds to type-egalitarian judgements that employ the predicate 'a positive cause of' (C^3_{30}), the second sentence seems to correspond to type-egalitarian judgments that employ the predicate 'a negative cause of' (C^3_{32}), and the third sentence seems to correspond to type-egalitarian judgements that employ the predicate 'is causally neutral' (C^3_{34}).

In order to remove all ambiguity I also assume that these sentences are meant to be token-egalitarian causal judgements. Once that assumption is made it is obvious that *all* judgements find their place in the theoretical framework of Lewis's counterfactual token-event analysis of causation. This raises an interesting question: *how can a paradigmatically monist analysis do that?*

The answer is that expressions such as '*a* prevented *b*' are translatable into expressions such as '*a* caused the absence (or omission) of *b*' and that the proponent of counterfactual theory also has the resources to *deny* that *a* is a cause of *b*. My argument at this point is that one doesn't call it 'pluralism' if one encounters those distinctions in one theory, so why should it be called 'pluralism' if very similar distinctions are encountered in another theory? It doesn't seem to be a 'pluralist' accomplishment to distinguish positive from negative causes and to have a separate concept for the denial of causal relevance. Any theory that didn't make those distinctions would fall short of standards set by *monist* theories such as the counterfactual analysis.

A possible objection at this point would be to argue that I haven't shown that 'intramural pluralism' isn't pluralism but that I have shown that a (supposedly) monist position turns out to be a pluralist position. I agree that my argument would also work in the opposite direction. But no matter in which direction one applies the argument, I have shown that the gap between these positions is much closer than the label 'intramural *pluralism*' suggests.

Obviously, the comparison I have just made with colloquial judgements is still incomplete: I haven't drawn parallels to C^3_{36} (mixed cause) and C^3_{38} (causally relevant).

So, do we have any colloquial expressions corresponding to judgements of that sort? If one considers the concept of a mixed cause, it doesn't seem as if there is such a direct correspondence. But I think that Eells's suggestion of introducing the notion of a mixed cause is a reaction to a problem that is very similar to a difficulty I discussed already.

The problem motivating Eells's account was brought up by Cartwright who considered the possibility that two causal factors may interact:

> For example, ingesting an acid poison may cause death; so too may the ingestion of an alkali poison. But ingesting both may have no effect at all on survival.
> In this case, it seems, there are three causal truths: (1) ingesting acid without ingesting alkali causes death; (2) ingesting alkali without ingesting acid causes death; and (3) ingesting both alkali and acid does not cause death. (Cartwright, 1979, p. 428)

Though neither Cartwright nor Eells makes that connection, this example is strongly reminiscent of an issue discussed in Sect. 2.1: the problem that a single factor is usually insufficient to bring about the effect and the resulting insight that what is usually called 'the cause' is an arbitrary selection from a conjunction of factors collectively causing the effect (Millian arbitrary selection principle). A difference is of course that Mill discusses the problem in terms of invariable regularity, i.e. necessitation, rather than imperfect correlation, i.e. probability change. But the problem is in essence the same. If one discusses the problem in terms of necessitation, the complaint is that the single factor is insufficient. If one discusses the problem in terms of probability, the complaint is that the single factor alone cannot guarantee a certain probability change. In other words, a single factor is an insufficient condition for a probability change. Cartwright's suggested solution also echoes Mill: if the causal relation doesn't hold between a single factor and an effect, then the causal relation holds between a conjunction of factors and an effect.

Eells calls Cartwright's strategy the 'combined-factors approach' (Eells, 1991, p. 130) and contrasts it with his own account, i.e. the 'revised contexts approach' (Eells, 1991, p. 131). The notion of a mixed cause offers an alternative solution to the same problem. Eells suggests that

> [a]mong individuals who have just ingested an alkali (acid) poison, ingesting an acid (alkali) poison is causally *negative* for death; among individuals who have not just ingested an alkali (acid) poison, ingesting an acid (alkali) poison is causally *positive* for death; and in the combined population, ingesting an acid (or alkali) poison is causally *mixed* for death. (Eells, 1991, p. 130, italics in original)

The obvious difference between this and the previous strategy is that it introduces a further notion of cause that doesn't demand that a cause either raises or lowers the probability of its effect. The more subtle difference between this and the previous strategy is that it makes a different assumption about the cause. According to the former strategy, the cause is a combination of factors; according to the latter strategy, the cause is a single factor.

It would seem that Cartwright's suggestion is the more conservative. It allows one to solve the problem by means of a familiar conceptual framework. Eells's strategy, in contrast, introduces a concept of cause that doesn't seem to correspond to any colloquially used causal expression. Hence, it wouldn't be right to conclude that the Eells-terminology simply mirrors the usual conceptual distinction that is already found in the colloquial discourse. There is a difference between the Eells-terminology and the colloquial discourse, although I don't think that this undermines my argument against Hitchcock's theory being pluralistic.

In order to get a comprehensive overview of 'intramural pluralism', let me now proceed to a discussion of the concepts of net effect and component effect. As mentioned above, Hitchcock thinks that there is a risk of obscuring this distinction if a stronger pluralist thesis than 'intramural pluralism' is adopted.

So, what is this distinction? Hitchcock speaks of 'path-specific causation' (Hitchcock, 2012, Sect. 2.9). The idea is to introduce causal concepts that allow us to differentiate between various relations that may hold between one and the same pair of variables.

The definitions of net and component effect were quoted in Sect. 2.4. Expressed in more natural terms, the notion of a net effect forbids the holding of fixed causal intermediates between cause and effect, while the notion of a component effect requires the holding of fixed causal intermediates on causal routes *other* than the route which one is examining. That is the crucial difference.

The paradigm example that motivates the distinction between net and component effects is a causal scenario first discussed by Germund Hesslow (1976): Contraceptive pills (or birth control pills) have several effects. On one hand, there is the pill's intended effect, i.e. to reduce the probability of pregnancy. On the other hand, the pill has several side effects, among them the effect of inducing thrombosis, i.e. raising the probability of thrombosis. Taken together with the information that pregnancy also increases the probability of thrombosis, this leads to the seemingly paradoxical conclusion that the pill causes *and* prevents thrombosis (by virtue of preventing pregnancy).[9] However, the puzzle resolves quickly once one distinguishes the two component effects (raising the probability of thrombosis directly; lowering the probability of thrombosis indirectly) and the net effect (raising or lowering the probability of thrombosis on the whole).

Hitchcock (2001, pp. 208 ff.) discusses the distinction between net and component effect in terms of hypothetical interventions, i.e. he imagines how one might hypothetically intervene on variables or fix variables in various combinations. Yet I believe it is more instructive to consider how one reveals such relations without hypothetical interventions. In practice one needs to condition on different populations in order to identify net and component effects. If one wants to know the component effect that induces thrombosis directly, i.e. without the causal intermediate of being pregnant, one has to compare two populations: non-pregnant women who take the pill and non-pregnant women who do not take the pill. Such a selection 'fixes' causal intermediates on the other causal route. The other component effect, i.e. the component effect that has pregnancy as causal intermediate, requires a

9 Note that the distinction between net and component effect is missing in Hesslow (1976). Hesslow concludes that this example demonstrates how a cause (taking the pill) may lower the probability of its effect (suffering from thrombosis). That there are two concepts of effect isn't a conclusion in that article.

comparison between other populations: women who conceive despite taking the pill and women who conceive without taking the pill. This 'fixes' causal intermediates on the other causal route. The assessment of the net effect requires still another comparison: those women who take the pill (regardless of whether they conceive) and those women who do not take the pill (regardless of whether they conceive). This selection does *not* 'fix' causal intermediates.[10]

The idea behind the notion of a component effect is to identify the more complex causal structure that underlies the net effect. There are distinct causal mechanisms bringing about thrombosis. The concept of a component effect helps one articulate that knowledge. In other words, *to hypothesise that there is more than one component effect is tantamount to the hypothesis that there is more than one causal path running from the cause-variable to the effect-variable*. This is in essence the idea behind a *plurality of causes* thesis. Recall the Millian plurality of causes principle from Sect. 2.1: it states that any type effect may have a plurality of type causes. The type effect death, for instance, has the type causes stabbing, suffocating, poisoning, drowning, and so on. Here the idea is quite similar: thrombosis has more than one cause. The only additional information is that one cause of thrombosis (the pill) is also the preventer of another cause of thrombosis (pregnancy). This is a symmetrical thesis to Mill's plurality of causes thesis: it states that a cause can have more than one effect. There are interconnections between Mill's idea and the terminology of net and component effect.

There are, however, also some differences. The notion of a component effect requires imagining a *population*, such that it is true to make a causal claim about a component effect of that population. The Millian plurality of causes principle is *not* restricted in this sense. It is an unrestricted principle that applies whenever and wherever the causal relata under consideration are types. Furthermore, the notion of a component effect requires imagining an accompanying net effect: A certain factor such as the intake of the pill may have a certain net effect on thrombosis that is brought about by several component effects. But there may be other causes of thrombosis in the same population, e.g. dehydration, that need not be component effects of the net effect that the pill has on thrombosis. So, the notion of a component effect articulates a specific plurality of causes: *the plurality of causes that are component effects of a certain net effect in a certain population.*

If one considers the problem from this perspective, it allows us to identify a peculiarity: it seems as if one has to assign sets of ternary relations to one predicate and sets of quaternary relations to the other predicate (provided that one adheres to a consistent translation policy):

C^3_{40}: ① has a net effect on ② in the population ③
C^4_{42}: ① has a comp. effect on ② in the population ③ given the sub-population ④

[10] In a more exhaustive account one would perhaps also try ruling out other interferences such as the indirect effects of other contraceptives, e.g. condoms.

This is an interesting result, since it challenges the idea of subsumption which was introduced earlier. I originally encountered a minor difficulty when applying that idea: in order to avoid the trivial failure of my account I suggested, for instance, not discriminating between events and processes (Sect. 3.2). This time, the problem is more profound. How should one subsume concepts of different arity into one another?

Suppose (for the sake of argument) that C_{40}^3 is the more general concept: how should a concept that has ternary relations as its extension subsume a concept that has quaternary relations as its extension? (It is impossible that the set of quaternary relations is a subset of the set of ternary relations.) Or, if we suppose that C_{42}^4 is more general: how should a concept that has quaternary relations as its extension subsume a concept that has ternary relations as its extension? (It is also impossible that the set of quaternary relations is a subset of the set of ternary relations.) Or, if both C_{40}^3 and C_{42}^4 are supposed to be subsumed by a third concept: what would that subsuming concept be? It would appear that such a subsuming concept would have to be a predicate of *indeterminate* degree, i.e. a predicate that is both of degree 3 and degree 4. The extension of such a predicate would have to be a set of ternary and quaternary relations.

Once one arrives at this conclusion, it is natural to propose an account of arity pluralism in addition to my account of (plain) pluralism. A theory exhibits arity pluralism when it postulates causal concepts of different arity.

I suggest that this idea may even shine further light on the thesis that causal concepts are only connected through family resemblance. If it is true that causal concepts vary not only according to their extension, where it is assumed that the elements in these sets are homogeneous n-tuples, i.e. n takes the same value, but also according to their arity, i.e. n takes different values, it becomes increasingly difficult to explicate how an abstract general notion of cause might be held together. One could say that such a concept is of indeterminate degree and not projectable as a translation into $\mathfrak{L}$.

However, I would not see this as an argument against the existence of such a notion of cause. $\mathfrak{L}$ is a rigid, limited and artificial language. Natural languages are much more flexible in dealing with abstract notions. One could also argue that this shows how resourceful natural languages are.

3.4 Conclusion and Outlook

I hope to have shown how extensional analysis can be used as a tool for bringing order to the idea of conceptual pluralism. This approach allows one to give an account of conceptual pluralism that goes beyond the mere truism that there are many concepts of cause. It also permits the observation that there is plurality along two dimensions: (plain) pluralism and arity pluralism.

Note, however, that—strictly speaking—I am also arguing for the irrelevance of these distinctions. On the one hand, it is important to have a clear view of what a reasonable account of 'monism' and 'pluralism' might be; on the other, I shall argue that positioning oneself on the monism-pluralism spectrum is often (though not always) of little consequence.

Although this may sound contradictory, the conclusion that one need not care much about whether one subscribes to monism or pluralism relies of course on a prior notion of monism and pluralism. In the following chapters this account sometimes serves as an analytic tool, but more often as a background against which I discuss conceptual plurality.

An account of pluralism is useful when one wants to point out that some distinctions are no more than words, as I will argue against Mayr (Chap. 4). In other cases, conceptual distinctions may be useful and yet not be instances of pluralism, since the concepts are easily subsumable to a master concept (Chap. 5). In yet other cases one can argue that a distinction is a true pluralist account but unrelated to the problems it is supposed to solve (Chap. 6). And finally one may still rank theories with respect to how well they solve specific tasks so that pluralism is unlikely to be the answer (Chap. 7). This leaves some room for epistemological-methodological pluralism, as outlined in Sect. 1.2, but it isn't a strong argument for any far-reaching pluralist claims.

One may conceive of this as an argument against pluralism. However, I don't think of this as a dogmatic claim. It is in principle possible to refute this claim by providing examples in which a meaningful pluralist conceptual distinction contributes to the solution of some still unknown problem. What I am warning against is throwing around conceptual distinctions that obscure the real problem in need of discussion.

References

Anscombe, G. E. M. (1993). Causality and determination. In E. Sosa & M. Tooley (Eds.), *Causation* (pp. 88–104). New York: Oxford University Press. (Reprinted from *Causality and determination. An inaugural lecture*, 1971, Cambridge University Press)

Boghossian, P. A. (2006). *Fear of knowledge. Against relativism and constructivism*. Oxford: Clarendon Press.

Cartwright, N. (1979). Causal laws and effective strategies. *Noûs*, *13*(4), 419–437. https://doi.org/10.2307/2215337

Cartwright, N. (2004). Causation: one word, many things. *Philosophy of Science*, *71*(5), 805–820. https://doi.org/10.1086/426771

Dowe, P. (2000). *Physical causation*. Cambridge: Cambridge University Press.

Eells, E. (1991). *Probabilistic causality*. Cambridge: Cambridge University Press.

Fitting, M. (2015). Intensional logic. In E. N. Zalta (Ed.), *The Stanford encyclopedia of philosophy* (Summer 2015 ed.). Metaphysics Research Lab, Stanford University. Retrieved from http://plato.stanford.edu/archives/sum2015/entries/logic-intensional/ (First published Thu Jul 6, 2006; substantive revision Thu Apr 2, 2015)

Frege, G. (1948). Sense and reference. *The Philosophical Review*, *57*(3), 209–230. https://doi.org/10.2307/2181485

Frege, G. (1986). Über Sinn und Bedeutung. In G. Patzig (Ed.), *Funktion, Begriff, Bedeutung. Fünf logische Studien* (6th ed.). Göttingen: Vandenhoek & Ruprecht.

Hall, N. (2004). Two concepts of causation. In J. Collins, N. Hall, & L. A. Paul (Eds.), *Causation and counterfactuals* (pp. 225–276). Cambridge, MA and London, England: MIT Press.
Hart, H. A. L., & Honoré, T. (1985). *Causation in the law* (2nd ed.). Oxford: Clarendon Press. (Original work published 1959)
Hesslow, G. (1976). Two notes on the probabilistic approach to causality. *Philosophy of Science, 43*(2), 290–292. https://doi.org/10.1086/288684
Hitchcock, C. (2001). A tale of two effects. *The Philosophical Review, 110*(3), 361–396. https://doi.org/10.2307/2693649
Hitchcock, C. (2007). How to be a causal pluralist. In P. Machamer & G. Wolters (Eds.), *Thinking about causes* (pp. 200–221). Pittsburgh, PA: University of Pittsburgh Press.
Hitchcock, C. (2012). Probabilistic causation. In E. N. Zalta (Ed.), *The Stanford encyclopedia of philosophy* (Winter 2012 ed.). Metaphysics Research Lab, Stanford University. Retrieved from http://plato.stanford.edu/archives/win2012/entries/causation-probabilistic/ (First published Fri Jul 11, 1997; substantive revision Sun Mar 21, 2010)
Hoyningen-Huene, P. (1998). *Formale Logik. Eine philosophische Einführung*. Stuttgart: Reclam.
Lewis, D. (2004). Causation as influence. In J. Collins, N. Hall, & L. A. Paul (Eds.), *Causation and counterfactuals* (pp. 75–106). Cambridge, MA and London, England: MIT Press. (Reprinted from *Journal of Philosophy*, *97*(4), 182–197 (2000); minor revisions have been made for consistency)
Machamer, P. (2004). Activities and causation: the metaphysics and epistemology of mechanisms. *International Studies in the Philosophy of Science, 18*, 27–39. https://doi.org/10.1080/02698590412331289242
Machamer, P., Darden, L., & Craver, C. F. (2000). Thinking about mechanisms. *Philosophy of Science, 67*(1), 1–25. https://doi.org/10.1086/392759
Mates, B. (1972). *Elementary logic* (2nd ed.). New York: Oxford University Press.
Wittgenstein, L. (1984). *Tractatus logico-philosophicus. Tagebücher 1914–1916. Philosophische Untersuchungen* (Vol. 1). Frankfurt am Main: Suhrkamp.
Woodward, J. (2003). *Making things happen. A theory of causal explanation*. New York: Oxford University Press.

Chapter 4
The Concepts of Ultimate and Proximate Cause

Abstract This chapter examines Ernst Mayr's distinction between ultimate and proximate causes. I first critically reconstruct the terminology and point out inconsistencies. I then discuss known objections to Mayr by Richard C. Francis and André Ariew. I show that despite its inconsistencies Mayr's account can be easily defended against Francis's objection that ultimate explanations are just functional explanations. Defending Mayr against Ariew's objection is also possible, since Ariew's argument relies on a false dichotomy between dynamical and statistical explanations. This chapter concludes with a discussion of why I nevertheless reject Mayr's terminology. The primary problem is that it collapses into a naive account of explanatory relativity that is incompatible with reasoning in biology.

Keywords Bird migration · Co-evolution · Dominant cause · Dynamical explanation · Evo-devo · Explanatory relativity · Extended evolutionary synthesis · Feedback · Functional explanation · Proximate cause · Statistical explanation · Ultimate cause

The essay *Cause and Effect in Biology* is famous for Ernst Mayr's promotion of the distinction between ultimate and proximate causes in biology. As was pointed out by John Beatty (1994), Mayr's defence of that terminology didn't serve a mere theoretical purpose. It was also a distinction that Mayr utilised as a means of securing his naturalist approach to biology against a perceived threat from the emerging field of molecular biology, which had at the time just succeeded in uncovering the molecular structure of DNA (Watson & Crick, 1953).

My aim in this chapter is to analyse Mayr's account of ultimate and proximate causes in a way that separates Mayr's rhetoric from the naked theoretical account. I shall argue that in pursuing a theory of ultimate and proximate causes Mayr is trapped between two objectives that are in tension with each other: on the one hand, he is denying that there is more than one sort of causality; on the other, he is maintaining that there are two somehow different sorts of causality.

The sort of causality he rejected was the Aristotelian-style final cause, as advocated by some vitalist biologists right up to the beginning of the twentieth

K. Ehrenstein, *Causal Pluralism in the Life Sciences*,
History, Philosophy and Theory of the Life Sciences 25,
https://doi.org/10.1007/978-3-030-87942-6_4

century. Even though vitalism no longer had any serious supporters by 1961, Mayr argues against vitalism in his essay and attempts to explain how to settle the dispute between those authors who believed in vitalism and those authors who favoured mechanist explanations of life (Mayr, 1961, p. 1501).

His account of how to replace the vitalist explanation of purpose with a mechanist one is accompanied by an illustration of differences between so-called *how* and *why* questions, and by a statement of the distinction between ultimate and proximate causes. Since Mayr is arguing against vitalism, it is evident that ultimate and proximate causes have to be mechanist causes or, in Aristotelian terminology, efficient causes. What Mayr doesn't admit to himself, since it would undermine his rhetoric against molecular biology, is that the ultimate-proximate distinction has to be a distinction in name only.

4.1 Precursors of Mayr's Ultimate-Proximate Distinction

Ernst Mayr did not invent the terminology of 'proximate' and 'ultimate' causes. He traces the origin of this ultimate-proximate terminology back to Herbert Spencer and George Romanes; however, he refers to John R. Baker as

> the first author to distinguish clearly between ultimate causes responsible for the evolution of a given genetic program (selection) and proximate causes responsible, so to speak, for the release of the stored genetic information in response to current environmental stimuli (Mayr, 1982, p. 68).

More precisely, Baker (who discusses the evolution of the breeding season, a topic closely related to the problem of bird migration later discussed by Mayr) distinguished *three* types of cause, not just two. According to Baker, one has to differentiate between 'ultimate cause', 'proximate cause' and 'artificial cause' (Baker, 1938, p. 162). Baker describes the three notions of cause as follows:

> Animals have evolved the capacity to respond to certain stimuli by breeding. In cold and temperate climates it is usually clear that the season adopted allows the young to grow up in favourable climatic conditions, and one may say that in a sense these conditions are the *ultimate cause* of the breeding season being at that particular time. There is, of course, no reason to suppose that the particular environmental conditions favourable to the young are necessarily the one or ones which constitute the *proximate cause* and stimulate the parents to reproduce. Thus abundance of insect food for the young might be the ultimate, and length of day the proximate cause of a breeding season. Agencies which start and stop reproduction but which do not operate under natural conditions of existence may be called *artificial causes*. (Baker, 1938, p. 162, italics added)[1]

According to this account, 'ultimate cause' refers to conditions such as the cold and temperate climates or the abundance of food under which the breeding

[1] This quote from Baker and the following quote from Thomson are also discussed in Beatty (1994) where you can find a second opinion on how to interpret these sources.

season evolved, while 'proximate cause' refers to a sort of disturbance or change in conditions such as the shortening and lengthening of the day. Hence 'proximate cause' appears to be an inegalitarian concept of cause. The same appears to be true of artificial causes. The 'artificial cause' is a human interference or disturbance, while the proximate cause is a natural interference or disturbance.

Classifying Baker's notion of an ultimate cause presents a greater problem. It appears that the examples given by Baker are no *arbitrary* selections. Baker's notion of an ultimate cause also seems to be an inegalitarian concept of cause. Conditions such as the climate or the food supply appear to be salient for some reason. Yet it seems less clear why they are chosen as salient causes. When he speaks about the 'cold climate', is it relevant that there are *changes* such as a change from warm to cold climate, or is the crucial point the fact that cold weather *recurs regularly* in a cold climate? A similar question concerns the 'abundance of food'. 'Abundance of food' may also be a factor that varies according to season but that recurs regularly. So, what is relevant? That there are long-term *changes* (in climate or food supply), or that there are regular recurrences that are *stable* over a long period of time? It could be that 'ultimate causes' are *not* chosen for the reason that they are disturbances or interferences. However, those differences are also not pointing towards the idea of an egalitarian concept of cause. One might therefore conjecture that 'ultimate cause' is probably also a dominant or inegalitarian concept of cause, for the reason that this would be consistent with the way 'proximate causes' are presented.

Another author who made a similar distinction was A. Landsborough Thomson (1926, Chap. XVI). His view of bird migration is the central topic of an essay Mayr wrote together with Wilhelm Meise (Mayr & Meise, 1930). Thomson suggested that there is a 'dual aspect' when one speaks about the causes of bird migration (Thomson, 1924, p. 639). He proposed that

> [t]he ultimate cause of migration must surely lie in the existence of the inborn habit and in the nature of the forces in the far past which gave it origin. In the second place there must be immediate stimuli, periodically recurring, which evoke the habit to active expression each autumn and each spring. (Thomson, 1924, p. 639)

This explanation of the ultimate-proximate distinction differs from Baker's account with respect to the emphasis that is given to the temporal dimension. Thomson contrasts 'the forces in the far past' with 'immediate stimuli'. Attention to time isn't absent from Baker's account, but it is less prominent.

In Beatty's analysis of Mayr's usage of the ultimate-proximate terminology, this temporal criterion is presented as if it were the cornerstone of the distinction (Beatty, 1994, p. 334). Yet is also true that Mayr credits neither Baker nor Thomson in *Cause and Effect in Biology*. This only happens elsewhere, for instance in the passage from *The Growth of Biological Thought* quoted above, which acknowledges Baker (though not Thomson).

This deserves attention, because Beatty's discussion of Mayr's account is very generous in the sense that he glosses over gaps in Mayr's reasoning, one of which is Mayr's inconsistent application of the temporal criterion (as will be seen in the discussion below). Beatty instead focuses his cautious criticism on the extent of

Mayr's influence, remarking that 'any more than minor disagreement with [Mayr's] insistence on the proximate/ultimate distinction would be heretical' (Beatty, 1994, p. 352). Considering how heated some of Mayr's exchanges with fellow academics were (Sect. 4.6), this might have been a wise decision (at that time).

4.2 Dividing Biology According to Question Types

In *Cause and Effect in Biology* Mayr explains the difference between ultimate and proximate causes by first introducing a distinction between *why* and *how* questions. While *how* questions are supposed to aim at proximate causes, *why* questions are supposed to aim at ultimate causes.

Mayr maintains that the typical research questions in functional biology (which roughly corresponds to physiology and molecular biology) are *how* questions, i.e. questions demanding an answer in terms of proximate causation:

> The functional biologist is vitally concerned with the operation and interaction of structural elements, from molecules up to organs and whole individuals. His ever-repeated question is "How?" How does something operate, how does it function? (Mayr, 1961, p. 1502)

By contrast, Mayr characterises the questions which recur in evolutionary biology as *why* questions, i.e. questions demanding an answer in terms of ultimate causation:

> Why did the warbler on my summer place in New Hampshire start his southward migration on the night of the 25th of August? (Mayr, 1961, p. 1502)

The general idea that researchers in different sub-disciplines of biology might pose different questions is fair enough. Chemists don't pose the same questions that linguists do. Why shouldn't the same apply to differently specialised groups within a discipline? This isn't anything that has to be called into question.

What is worrisome is that Mayr makes it look as if it were a matter of which interrogative is employed. This is clearly not the case, as Brett Calcott (2013, p. 769) has demonstrated by providing counterexamples that switch 'how' and 'why.' However, even more disturbing is that one doesn't even have to search for counterexamples. Mayr provides all the evidence against that terminology while explaining it.

According to Mayr, *why* questions need to be disambiguated: he distinguishes between *why* questions in the sense of 'how come?' and in the sense of 'what for?' (Mayr, 1961, p. 1502). *Why* questions, as Mayr conceives them, are meant to be a matter of purpose. He distinguishes those questions that aim at a mechanically explicable purpose (teleonomy) from those questions that aim at a mechanically inexplicable purpose (teleology). The purpose behind the warbler's behaviour (although only in hindsight) is to avoid death by starvation. This is in principle explicable as an outcome of evolution, i.e. as the effect of natural selection on the warbler's ancestors. This would be an answer to the *how-come* question. One answer to the *what-for* question might be that God created the warbler for some divine purpose. Another answer—and Mayr's favoured reply—would be that there is no purpose, since evolution (as a whole) has no purpose. *What-for* questions are therefore not (or, since Darwin, no longer) a subject of biological enquiry. It should be clear that, once *why* questions are disambiguated into *how-come* and *what-for* questions, the interrogative 'why' is no longer relevant. The distinction would be between *how-does* questions (*how* questions), *how-come* questions (*why* questions), and *what-for* questions (*why* questions). Obviously, nothing hinges on the usage of the word '*why*.'

It also doesn't seem to be of great importance to use the interrogative '*how*', either. An alternative formulation Mayr uses for his question of why the warbler left New Hampshire on the night of the 25th of August is '[w]hat is the cause of bird migration?' (Mayr, 1961, p. 1502). Had he been serious about the difference between *how* and *why* questions, he should have asked, 'How does the mechanism operate that triggers bird migration?' This would at least have illustrated his thesis that one can formulate *how* and *why* questions about the same phenomenon, although this still could not have shown that it is the interrogatives that make the difference.

Some still take the distinction between *why* and *how* questions for granted, maintaining that '[b]iologists will always require different answers to how and why questions' (Laland et al., 2011, p. 1516). David Haig thinks '[a] distinction between "How (incorporating how come)?" and "What for?" remains useful' (Haig, 2013, p. 785). But there are also other voices; e.g. Ward B. Watt, who states that the distinction is 'illusory', since *why* questions transform into *how* questions (Watt, 2013).

I suspect that this apparent disagreement may not run as deep as it seems to, if one takes the very plausible view that researchers who are specialised in different areas pose different questions, rather than the less convincing view that the labels 'how' and 'why' are part of a substantial explanation of the alleged differences between proximate and ultimate causes.

4.3 New Hampshire on the Night of the 25th of August

The central example Mayr uses to illustrate the difference between ultimate and proximate causes is bird migration. As mentioned in Sect. 4.2, he uses two formulations—a *what* (!) and a *why* question—to introduce the topic. When he goes on to claim that he 'can list four equally legitimate causes for this migration' it almost appears as if he thinks of those questions as interchangeable, although this would of course contradict his thesis that one has to differentiate between *how* and *why* questions (Mayr, 1961, p. 1502).

What does 'this migration' mean in Mayr's claim? Is he referring to the particular event of the warbler leaving New Hampshire on the night of the 25th of August or to the more general phenomenon of bird migration? Although one may have expected that Mayr wanted to refer to bird migration in general, it appears to be a more natural reading to assume that he is concerned with a particular token event on the night of the 25th of August.

Mayr claims that this event has to be explained by four different causes. There are two hypotheses citing *ultimate* causes:

1) *An ecological cause.* The warbler, being an insect eater, must migrate, because it would starve to death if it should try to winter in New Hampshire.
2) *A genetic cause.* The warbler has acquired a genetic constitution in the course of the evolutionary history of its species which induces it to respond appropriately to the proper stimuli from the environment. On the other hand, the screech owl, nesting right next to it, lacks this constitution and does not respond to these stimuli. As a result, it is sedentary. (Mayr, 1961, p. 1502, italics in original)

And there are two answers citing *proximate* causes:

3) *An intrinsic physiological cause.* The warbler flew south because its migration is tied in with photoperiodicity. It responds to the decrease in day length and is ready to migrate as soon as the number of hours of daylight have dropped below a certain level.
4) *An extrinsic physiological cause.* Finally, the warbler migrated on the 25th of August because a cold air mass, with northerly winds, passed over our area on that day. The sudden drop in temperature and the associated weather conditions affected the bird, already in a general physiological readiness for migration, so that it actually took off on that particular day. (Mayr, 1961, p. 1503, italics in original)

Note that these answers further contribute to the ambiguity of Mayr's account. What are the causal relata? It would appear that they could be events, but also properties or dispositions. It is easy to see that the proximate causes are described as concrete token events, i.e. shortening of daylight and drop in temperature. These changes in the environment trigger the dispositions of 'photoperiodicity' and being 'in a general physiological readiness'.

The aspect that stands out from the ultimate causes is the properties or dispositions, i.e. being an insect eater and possessing a genetic constitution. That, however, isn't all. The idea of token-event causation still appears to be in the background when Mayr speaks about the ultimate causes.

Mayr's formulation of the ecological cause refers to an event that would happen (starving to death) if the warbler didn't migrate. It is of course strange that this refers to a counterfactual event that would only manifest itself in the future. If this were the effect, it would result in an inverted temporal order in which the effect is prior to the cause. One would then be dealing with a final instead of an efficient cause.

When Mayr speaks of the genetic cause, token-event causation is still in the background. The property (or more specifically the disposition) of having a certain genetic constitution induces a response, i.e. a token event, that is triggered by the stimuli from the environment. This also results in a problem: if one were to apply Thomson's account that distinguishes the forces in the past from the immediate stimuli (Sect. 4.1), one would have to conclude that the genetic cause falls into the same category as the two physiological causes, i.e. one would have to think of the genetic cause as a proximate cause.[2]

In short, Mayr's example of bird migration raises more questions than it answers. One cannot safely deduce anything from this carelessly-stated example.

4.4 The Program Account of Ultimate and Proximate Causes

The example of bird migration serves as an illustration of what one might call 'Mayr's program account of ultimate and proximate causes'. This is Mayr's explanation of how observations by vitalists are explicable in mechanist terms. He returns to this issue in various other publications.

The key element of this account is the idea that it is the genetic programs that make a significant difference between proximate and ultimate causes. Mayr argued that

> [o]rganisms, in contrast to inanimate objects, have two different sets of causes *because organisms have a genetic program*. Proximate causes have to do with the decoding of the program of a given individual; evolutionary causes have to do with the changes of genetic programs through time, and with the reasons for these changes. (Mayr, 1982, p. 68, italics added)

This makes it much clearer what Mayr's criterion is; namely, the mode of interaction with genetic programs. The ultimate causes change genetic programs, the proximate causes execute genetic programs. Since causes should be spatiotemporally localised, one might ask where those programs are instantiated. For this question Mayr has a plain and simple answer. The material instantiation of genetic programs is the DNA:

> The [...] DNA code of every zygote (fertilized egg cell), which controls the development of the central and peripheral nervous systems, of the sense organs, of the hormones, of physiology and morphology, is the *program* for the behavior computer of this individual. (Mayr, 1961, p. 1504, italics in original)

Following this idea, one can recognise ultimate causes quite easily: they should be mutations, since they change DNA sequences; and they may also be meiosis and syngamy, since they rearrange DNA. The odd thing is that this expectation isn't met by Mayr, as will be become obvious further below.

Another strange consequence is that the criterion of how causes interact with the genetic material further undermines Thomson's suggestion that causes act on different timescales. This outcome of the analysis was probably not intended by Mayr, since he emphasised the temporal dimension:

> Every organism, whether individual or species, is the product of a long history, a history which indeed dates back more than 2000 million years. (Mayr, 1961, p. 1502)

---

[2] I am grateful to an anonymous reviewer for pointing out this consequence to me.

It should be evident that although many ultimate causes happened 2000 million years ago, there are also ultimate causes that are happening in the present, according to Mayr's program account of ultimate and proximate causes. As soon as one adopts that account the issue of timescale becomes secondary. Why would Mayr use this second criterion if categorisation by timescale sorts causes differently than categorisation by mode of interaction with the genetic code? It appears that part of the explanation is Mayr's discussion of vitalism alongside this issue.

Mayr's account of a genetic program isn't exhausted by saying that DNA instantiates genetic programs. While DNA is one possible instantiation of programs, Mayr describes programs, i.e. genetic programs and programs in general, as

> coded or prearranged information that controls a process (or behavior) leading it toward a given end. (Mayr, 1974, p. 102, italics removed)

According to this view, a program is something that is holding information and leading to an end. Paradigm examples that are meant to fall under this definition are computer programs. The inclusion of computer programs allows Mayr to draw an analogy between computers and organisms. Mayr argues that a computer can act as if it had a purpose, although it hasn't, because it has been programmed to perform certain operations. He suggested that the same is true of biological organisms that obey a genetic program:

> A bird that starts its migration, an insect that selects its host plant, an animal that avoids a predator, a male that displays to a female—they all act purposefully because they have been programmed to do so. (Mayr, 1961, pp. 1503 f.)

The advantage of explaining purposeful behaviour in living things by analogy with computer programs is that a genetic program—understood as a strand of DNA—is a metaphysically non-obscure entity, the operations of which are explicable in terms of chemistry and physics. Invoking the notion of a genetic program is an attempt to reconcile 'a mechanistic interpretation of natural processes' with 'the seemingly purposive sequences of events in organic growth, in reproduction, and in animal behavior' (Mayr, 1961, p. 1503).

As mentioned, Mayr's essay is set against the background of vitalist biology, although vitalism had already been rejected at the time when *Cause and Effect in Biology* was written. It was a belief among vitalist biologists that there is a certain 'vitalistic or autonomous factor' (Driesch, 1908, p. 143) that guides embryonic development. Vitalists such as Hans Driesch spoke of '*a true element of nature*' (Driesch, 1908, p. 142, italics in original), which was supposed to account for those aspects of embryonic development that he believed to be inexplicable in terms of physics or chemistry. Driesch—honouring Aristotle—called this factor 'entelechy' (Driesch, 1908, pp. 143 ff.). He believed he had shown that '[n]o kind of causality based upon the constellations of single physical and chemical acts can account for organic individual development' (Driesch, 1908, p. 142).

Mayr objected to vitalism on the basis that it would offer no real explanation, merely replacing the 'unknown' with the 'unknown' (Mayr, 1961, p. 1503). But Mayr also acknowledged that 'some of the underlying observations' by the vitalists 'are quite correct', despite noting that 'the supernaturalistic conclusions drawn from these observations are altogether misleading' (Mayr, 1961, p. 1503). From this perspective DNA replaces the vitalistic factor, thus bridging the gap between the vitalists' observations and the theoretical restrictions imposed by chemistry and physics.

That somehow a replacement is taking place can also been seen in Mayr's discussion of Aristotle. Mayr puts forward the idea that Aristotle's form-giving principle is in fact DNA. Only a few terminological modifications would be required in order to render Aristotle's writings intelligible in terms of modern biology:

> Aristotle saw with ex[t]raordinary clarity that it made no more sense to describe living organisms in terms of mere matter than to describe a house as a pile of bricks and mortar. Just as the blueprint used by the builder determines the form of a house, so does the *eidos* (in its Aristotelian definition) give the form to the developing organism and this *eidos* reflects the terminal *telos* of the full grown individual. [...] Much of Aristotle's discussion becomes remarkably modern if one inserts modern terms to replace obsolete 16th and 17th century vocabulary. There is, of course, one major difference between Aristotle's interpretation and the modern one. Aristotle could not actually see the form-giving principle (which, after all, was not fully understood until 1953) and assumed therefore that it had to be something immaterial. (Mayr, 1974, p. 111)

According to this view, life cannot be understood without paying attention to the form-giving principle, i.e. DNA. It would also appear that Mayr is still very interested in teleological causal explanations, although he avoided the term 'teleology' and substituted it with the term of art 'teleonomy' (Mayr, 1961, p. 1504). This expression is borrowed from Colin S. Pittendrigh who used it 'in order to emphasize that the recognition and description of end-directedness does not carry a commitment to Aristotelian teleology as an efficient causal principle' (Pittendrigh, 1958/1967, p. 394). Mayr applies the term 'teleonomy' to the behaviour of organisms, but also to processes such as growth when the process depends on a genetic program (Mayr, 1974, pp. 98 ff.; Mayr, 2004, pp. 51 ff.). He does not consider evolution as a whole, however, to be a teleological process. Mayr maintained that '[t]he word *purpose* is singularly inapplicable to evolutionary change' (Mayr, 1961, p. 1504, italics in original).

Mayr's discussion of vitalism may explain how the temporal criterion for distinguishing ultimate from proximate causes turns out to be less important on closer inspection. Of course, this would not yet remove the ambiguity from Mayr's use of bird migration as an example. One might still ask whether it is possible to apply the criterion of how causes interact with genetic programs consistently.

To reformulate the ecological cause, one would want to name an event that changes a genetic program and also to avoid referring to a future event as a cause. The ecological factor explicitly named by Mayr is food supply. However, food supply doesn't literally change a DNA sequence, although a lack of food supply can starve individuals to death, thereby destroying copies of certain genetic programs. Even though Mayr emphasises the change of programs (which evokes such ideas as mutation or sexual recombination of programs), he is more likely concerned with their selection. Hence, one might consider this reformulation:

> *Ecological cause (first revision)*: There are past events (the starving of warblers that didn't migrate) that have removed copies of genetic programs from the gene pool of the population.

Such a reinterpretation would shift the focus away from the particular event on the night of the 25th of August towards past events in the population to which the warbler belongs. Although this would change Mayr's literal wording in *Cause and Effect in Biology*, it would be warranted by his emphasis on so-called 'population thinking' elsewhere (e.g. Mayr, 1982, pp. 83 ff.). Note that this shift of focus doesn't mean that the idea of token causation is being discarded. My revision refers collectively to concrete token events of which each has a spatiotemporal location and is not a mere abstract event type that has no particular spatiotemporal location. This reconstruction of Mayr's thoughts would be supported by his remarks on uniqueness in biology. Later in *Cause and Effect in Biology* Mayr contrasts the phenomena in physics that exemplify recurring types with the phenomena in biology that he describes as unique occurrences on every organisational level (Mayr, 1961, p. 1505). This would lend support to the hypothesis that Mayr's example of bird migration can be reconstructed in terms of token-event causation.

However, my revision might still face criticism from a different angle: no starving event, even if they are all localised at a particular spatiotemporal position, is continuously linked to the token event on the night of the 25th of August. The warbler leaving New Hampshire on the night of the 25th of August inherited its genetic programs from those warblers that migrated and didn't die. Although one is no longer referring to some uninstantiated counterfactual future event, it has to be acknowledged that there is no spatiotemporal continuity between those alleged ecological causes and the particular warbler leaving New Hampshire at that specific time.

One might wonder whether this is the case with other ecological causes, too. Is it a general fact that ecological causes are not continuously linked to their effects? One can easily see that this is not the case. When an environment provides enough food to sustain certain individuals, this is also an ecological factor ensuring survival. Instead of pointing to all those warblers that didn't survive because they starved to death, one could also point to those warblers that did survive because they found enough food in those areas to which they migrated.

> *Ecological cause (second revision)*: There are past events (finding sustenance) that have resulted in a transmission of copies of genetic programs to the present.

There is a spatiotemporal continuous succession of events that links the warbler leaving New Hampshire on the night of the 25th of August to each of those successful foraging events of its ancestors. The only problem is that those ecological causes are *not* singled out by Mayr's program account of ultimate and proximate causes, since they don't change any genetic program. They might contribute to the successful spread of a program, but they are not the sort of causes which change programs.

Given that neither Mayr's original formulation of the ecological cause nor any obvious revision refers to literal changes in genetic programs, one has to question whether there is any coherent thought to be found in Mayr's account. One could try to test further ecological factors against Mayr's criterion of changing genetic programs. I, however, think the more reasonable conclusion is to accept that this is again a loosely formulated idea (like the distinction between *how* and *why* questions) that falls apart under scrutiny.

Mayr's failure to pay any attention to causes of variation, such as meiosis and syngamy or mutations, is a conspicuous lacuna in his illustration of ultimate causes of bird migration. Any standard account of modern evolutionary theory—Mayr (1982) included—conceives of evolution as an interplay of causes that create variation and causes that select individuals with a certain phenotype. Yet Mayr's example only refers to factors/events that select phenotypes (ecological cause), or factors/events that might be confused with proximate causes (genetic cause).

Looking at this mess, it would make much more sense to throw the terms 'ecology' and 'genetic' out of the window and to give an example of how variation occurred in the warbler's ancestors and an example of how the warbler's ancestors were selected (preferably one focussing on those selection events that result not in death but in survival). This would also allow the account to be realigned with Thomson's temporal criterion. Any event creating variation or selecting an individual can always be seen from two temporal perspectives: there are events in the present that create variation and select individuals, and there are events in the past that have created variation and have selected individuals. Whether a certain event is classified as 'ultimate' or 'proximate' would then depend on the effect that it causes. If it affects the same generation, it would be referred to as a 'proximate cause'; if it affects a subsequent generation, it would be referred to as an 'ultimate cause'. It should be clear that being an ultimate or proximate cause in this sense would be an *emphatically* relational property.[3] One and the same event could be a proximate cause of what happens in the present and a proximate cause of what happens in the future. The same would be true of past events that are proximate causes of effects that occurred in the past and ultimate causes of events that occur now.

Much of the confusion arises because Mayr is sidetracked by the issue of purpose and vitalism. In hindsight, it might look as if it would have been better if he had simply avoided that issue. That, however, would have drawn attention to the fact that 'ultimate' causes are ordinary (temporally distant) causes—a conclusion that would not have suited Mayr's interests. Beatty documents how Mayr (and researchers with a similar academic profile, such as Theodosius Dobzhansky or George Simpson) faced problems with the acquisition of funding due to the strong competition from molecular biology (Beatty, 1994, pp. 348 ff.).

---

[3] I say 'emphatically' because I conceive of all causal predicates as relational anyway. In this case, however, a reconstruction in terms of a one-place predicate would be even more inappropriate than usual.

While this is an explanation for Mayr's motives, it is of course no excuse for a lack of substantial argument. Mayr makes it look as if there was some profound difference between the two categories of cause when he asks, 'Is *cause* the same thing in functional and evolutionary biology?' (Mayr, 1961, p. 1502, italics in original). Yet the only true difference he is discussing is that between vitalist views of causation (which he rejects) and mechanist views (which he endorses). None of his talk about teleonomy or his computer analogy would be acceptable to him or to his audience without the assumption that there is nothing other than mechanist causation, i.e. efficient causes, in Aristotelian terminology. Consequently, 'ultimate' and 'proximate cause' can only refer to subsets of mechanist causes. How Mayr differentiates between these two sets of causes is unclear: a *strict* application of his program account of ultimate and proximate causes isn't possible.

Despite the fact that Mayr didn't succeed in providing a working criterion in his essay *Cause and Effect in Biology*, he was successful in the sense that many authors agree that he managed to drive a wedge between functional and evolutionary biology and that the gap he created needs to be bridged (e.g. Amundson, 2005; Laland et al., 2011; Calcott, 2013). Any effort to break Mayr's 'spell' on biological reasoning should involve the exposure of how his terminology fits into a nomenclature of general causal reasoning. This is what I shall now examine, before I continue with a discussion of objections that have been raised against Mayr's account of ultimate and proximate causes.

4.5 Extensional Analysis of 'Ultimate' and 'Proximate Cause'

The vagueness, inconsistency and ambiguity of Mayr's claim about bird migration leave it open to multiple interpretations. However, the historical predecessor accounts by Baker and Thomson, who appear to apply an inegalitarian token notion of cause (Sect. 4.1), and Mayr's presentation of the example as if he were speaking in terms of token-event causation (Sect. 4.3), allow one to suggest that his thesis is structured around the assumption that any biological event has four salient causes (relative to four different normal conditions[4]) in the sense that there is a (nonvacuous) true reading of

$$\forall z(B^1 z \rightarrow (\exists v C_1^3 v z f_1 \wedge \exists w C_1^3 w z f_2 \wedge \exists x C_1^3 x z f_3 \wedge \exists y C_1^3 y z f_4)) , \tag{4.1}$$

[4] Note that I do not think that it is plausible to have these four alternatives. My opinion is that one is not free to select just any normal condition, since the context determines which normal condition applies. For the sake of argument, however, I shall postpone this objection until Sect. 4.8.

where the interpretation $\mathfrak{I}$ is:

$\mathfrak{D}$: the set of token events and normal conditions
B^1: ① is a biological event
C_1^3: ① is the token cause of ② with respect to normal condition ③
f_1: the normal condition that abstracts from non-ecological causes
f_2: the normal condition that abstracts from non-genetic causes
f_3: the normal condition that abstracts from non-intrinsic-physiological causes
f_4: the normal condition that abstracts from non-extrinsic-physiological causes

One would then read the formula 4.1 as 'for any event z, if z is a biological event, then there is an event v that is the cause of event z with respect to the normal condition f_1 and there is an event w that is the cause of event z with respect to the normal condition f_2 and there is an event x that is the cause of event z with respect to the normal condition f_3 and there is an event y that is the cause of event z with respect to normal condition f_4'. Note that I do not claim that giving a name to those normal conditions explains what ultimate causes are. The normal conditions f_1 and f_2 are rather placeholders for whatever Mayr might have had in mind and cannot be properly reconstructed.

I suggest that such a formal rendering of Mayr's position even illustrates how the example of bird-migration goes astray when Mayr describes the genetic cause as if it were also a proximate cause. While the formula 4.1 states that there are salient causes relative to all four normal conditions, it doesn't imply or require that those causes have to be numerically distinct. According to this formalisation, there could be effects that have four numerically identical causes, i.e. there could be events that have four different roles relative to those four different normal conditions. This would of course not just solve the problem that genetic causes might be proximate causes, but also shed light on the problem that ecological and genetic causes might not be distinct.

Mayr would probably have wanted to deny the possibility of numerical identity. It would, however, be misleading to add an explicit condition that v, w, x and y have to be numerically distinct if Mayr's example doesn't fit the bill. Instead one should see formula 4.1 as an instrument that helps to clarify flaws in Mayr's reasoning without overshooting the mark. An important lesson from this formalisation is to see that Mayr's position results in strange consequences, but those consequences are *not contradictory in a strict sense*. That some proximate and ultimate causes may *not* be numerically distinct doesn't create a contradiction. The contradiction that remains is the tension between Thomson's temporal criterion and Mayr's programme account. Those principles are not working in harmony and it would be much better to abandon the programming metaphor in favour of the much clearer temporal criterion.

The other important lesson is of course that Mayr's talk of 'ultimate' and 'proximate causes' doesn't require any strong pluralist commitment. The reconstruction 4.1 works with the predicate C_1^3, which is the ordinary notion of salient token-event causation. One could replace this with the predicates

C_{51}^3: ① is the ultimate cause of ② with respect to normal condition ③

C^3_{53}: ① is the proximate cause of ② with respect to normal condition ③

and write

$$\forall z(B^1 z \rightarrow (\exists v C^3_{51} vzf_1 \wedge \exists w C^3_{51} wzf_2 \wedge \exists x C^3_{53} xzf_3 \wedge \exists y C^3_{53} yzf_4)) \,. \tag{4.2}$$

But all that would be demonstrated is that it is always possible to propose a causal concept with an extension that is merely a subset to a more general concept of cause. This reconstruction would still allow for some or even all ultimate and proximate causes to be numerically identical. As long as there is no argument that such concepts are not subsumable under an overarching general concept of cause (Anscombe or Hall, Sect. 3.2), no substantial pluralist thesis has been stated. The distinction is made in name only and it is not even clear how one sorts causes into one or the other category, since Mayr's program account does not group causes in the same way as Thomson's temporal criterion, which Mayr apparently would have wanted to use side by side.

4.6 Are Ultimate Causes Just Functional Explanations?

Some authors have flatly refused to accept Mayr's account of ultimate causes. In an essay critically examining the ultimate-proximate distinction Richard C. Francis concluded that '[t]here are no ultimate causes' (Francis, 1990, p. 413). This criticism in particular earned a rather harsh reply from Mayr who responded '[t]his shows how little [Francis] understands these terminologies' (Mayr, 1993, p. 93). Mayr attempted to clarify the situation by pointing out that he would now prefer to speak of 'evolutionary causation' instead of 'ultimate causation' (Mayr, 1993, p. 94).

This of course is a rather curious exchange: how can Mayr counter an objection to the existence of ultimate causes by responding that he has chosen to name them differently? And how can Francis doubt the existence of 'ultimate' (alias 'evolutionary') causes? There are, as my analysis in Sects. 4.3 and 4.4 has shown, reasons for criticising Mayr's account. What has yet to be seen is whether Francis raises another valid point against Mayr. One has to acknowledge that Mayr's defence is weak. His response shows that he is more concerned with—as one might describe it—fine-tuning the sense or intension of 'ultimate cause' than with responding to a problem that occurs independently of how one names the extension assigned to 'ultimate cause'. However, Francis's argument is also confused, since he does not distinguish between epistemic and ontological problems. It will be seen that it is for this reason that his attack on Mayr's account of ultimate causes does not succeed, even though Mayr's reply is unsatisfactory.

I shall begin my analysis with a brief synopsis of the background of Francis and Mayr's dispute. There is a long-standing debate concerning biological function. Participants in that debate have discussed which notion of biological function is preferable: the selectionist account (also known as the etiological account) or the systematic account (also known as the Cummins account) of biological functions.

An example of a biological function is that 'eyes are *for* seeing' (Perlman, 2010, p. 53, italics in original). According to the selectionist view, one has to look back at the evolutionary past if one wants to ascribe a function to the eyes. Proponents of that view would argue that eyes have been selected for seeing. In other words, there is a causal history that accounts for the function performed by eyes (Perlman, 2010, pp. 53 ff.). But according to the systematic view, there is no need to invoke the evolutionary past. The major proponent of this view, Robert Cummins, treats functions as a subset of dispositions or capacities, i.e. every function is a disposition but not every disposition is a function (Cummins, 1975, pp. 757 f.). If a function is a certain sort of disposition, it is possible to know the function of eyes without knowing anything about the evolutionary past (Cummins & Roth, 2010, p. 79).

Mayr, however, would have wanted to explain more than that. If, for instance, the warbler is disposed to leave New Hampshire at the end of summer, he would have wanted to give a causal explanation of that phenomenon in terms of ultimate causes. He would have wanted to explain how the warbler acquired the disposition (i.e. the 'function') that is triggered on the night of the 25th of August. Using Mayr's rhetoric, one could ask: how come the warbler has the disposition to leave at the end of summer?

Those who advocate the systematic view of functions would be sceptical about trying to answer this question. They would point out that one can recognise the disposition, i.e. function, of the warbler to leave on the night of the 25th of August independently of the evolutionary past. Defenders of the systematic view such as Cummins and Roth are not jumping to the conclusion that there are no such causes, but they are denying that such causes should be taken into consideration when one analyses biological functions (Cummins & Roth, 2010, pp. 75 f.).

A key problem—as should be obvious—is the question of how to get epistemic access to the causes in question: while Mayr believes that a program account of ultimate and proximate causes is needed in order to account for apparent purpose in biology, Cummins and Roth believe that no such account is needed and that one should confine oneself to the analysis of (systematic) functions. The detour to the historical past is in their view unnecessary.

At this point I want to return to the dispute between Francis and Mayr. Francis in fact has two interrelated objections to Mayr: one is that the extension of 'ultimate cause' is only a subset of the extension of 'phylogenetic cause'; the other is that there is no reliable methodology for recognising ultimate causes. One can deduce the first point from the way Francis criticises his opponents for refusing to speak of 'ontogenetic' and 'phylogenetic cause' instead of 'proximate' and 'ultimate cause':

> That these terms [Francis refers to 'ontogenetic' and 'phylogenetic cause'; K. E.] are not deemed adequate by sociobiologists [this is meant to include Mayr; K. E.] provides a clue to interpreting ultimate causes. Ultimate causes are not simply those events that left their imprint on the organism's genome; genetic drift, for example, is not usually accorded this status. (Francis, 1990, p. 405)

If one accepts Francis's interpretation, 'ultimate cause' is subsumable to 'phylogenetic cause': all ultimate causes are phylogenetic causes, but not every phylogenetic cause is also an ultimate cause. The conflict, it seems, is about how to define the extension of 'ultimate cause'. Which other events should fall under the notion of ultimate cause and which events should be excluded? In other words, should events such as genetic drift be regarded as events that are ultimate causes or should the extension of ultimate cause be restricted to those events in which natural selection takes place?

But then, if this is the conflict, why should someone opt for one solution or the other? Shouldn't clarifying the extensional difference suffice, instead of dismissing one pair of concepts in favour of another pair of concepts? So, why is there any need to choose? According to Francis, it is impossible to acquire knowledge of ultimate causes. He claims that Mayr and others fool themselves when they claim to investigate 'ultimate causes':

> Though explanations claiming to provide "ultimate causes" do usually refer to natural selection, it is typically rather indirectly. Ultimate explanations, in fact, usually do not concern causes, but rather, certain types of effects called "adaptations", the existence of which are inferred from functional analyses. Ultimate explanations are functional explanations. (Francis, 1990, p. 405)

In short, Francis believes that proponents of ultimate causation produce functional explanations instead of causal explanations. This of course is not the same as denying that there are ultimate causes. Francis refused to accept causal explanations in terms of ultimate causes because he believed that one couldn't show which of those explanations are true and which are false.

The problem, as Francis sees it, is that functions of the same type can have various causal histories and that a functional analysis cannot distinguish between these competing hypotheses. He argues that '[t]here are typically more than one means (mechanism) by which a function can be realized and which a functional account cannot discriminate' (Francis, 1990, p. 406). Further on he claims '[e]ven the most syncretic functional analysis, though, is no substitute for knowledge of etiology' (Francis, 1990, p. 412). This, however, is a methodological point. It doesn't show that there are no ultimate causes. It only shows that one is possibly ignorant of ultimate causes.

Mayr rejected Francis's thesis that functional and ultimate explanations are the same, and argued that Francis made an equivocal use of the word 'function'. Mayr then disambiguates 'function' into 'physiological processes' on the one hand and 'ecological (adapted) roles' on the other (Mayr, 1993, p. 94). Using the terminology introduced above, one could also say that Mayr distinguishes between 'systematic function' and 'selectionist function'.

It should be obvious that disambiguating the word 'function' is *no* adequate response to the objection that it is impossible to infer ultimate causes from functional analyses. Even if it were true that Francis conflates two notions of function, it would still be a problem that, according to Francis, one cannot have any knowledge about the causal history of a selectionist function. The epistemic problem would remain.

Mayr should have addressed this problem, too. A problem, as proponents of the systematic account of function see it, is that one cannot tell or that it is very difficult to tell whether a function, i.e. a disposition, is the result of a selection process or the result of some other process. Yet evolutionary biologists are frequently confronted with this kind of problem: they distinguish between homologous and non-homologous traits.

When biologists identify similar traits in different species, they usually direct their attention to three competing hypotheses that could explain the similarity in question: the trait in question can be a homoplasy, an analogy or a homology. A homoplasy is a shared trait that has *not* arisen because of common ancestry. The analogy is a special case of a homoplasy. If a trait is analogous, it has arisen independently in at least two species as a result of convergent natural selection. If a trait is homologous, it has arisen in a common ancestor of the (at minimum two) species under consideration. A typical textbook example for a homoplasy is the shared fusiform shape of aquatic animals, which is an adaptation for swimming (and therefore also an analogy). Ancestral relations show that the fusiform shape has arisen more than once in the course of evolution. A textbook example for a homologous trait is the similarity between the forelimbs of vertebrates. Even though there is a considerable variation between vertebrate forelimbs (e.g. wings, arms, forelegs, paddles or fins), all forelimbs share similarities with respect to the skeleton supporting these limbs. Ancestral relations show that these similarities are owed to common ancestry. Evolutionary biologists have developed a vast array of analytic tools for answering these kinds of question (Stearns & Hoekstra, 2005).

The question of whether a trait (or one could say 'function' or 'disposition') is homologous or non-homologous is answered by examining (i.e. reconstructing) phylogenetic relations. Another way of putting this would be to say that evolutionary biologists examine the *causal* history of traits. At minimum they distinguish between traits that are a result of natural selection and traits that are not a result of natural selection.

If Mayr had given a more comprehensive reply to Francis's attack, he could have pointed out that Francis is quite right when he claims that functional analysis doesn't tell us anything about ultimate causes. Functional analysis, however, is not what matters. It is phylogenetic analysis that provides the answers. Why Mayr didn't point out the importance of phylogenetic inference is unclear to me.

Mayr is at least partially to blame for giving a misleading impression in his essay *Cause and Effect in Biology*. He claims to explain which different methods biologists use when investigating either ultimate or proximate causes. But all that he actually does is to hint vaguely that evolutionary biologists cannot, for instance, rely on controlled and reproducible experiments. How evolutionary biologists overcome this problem is at best intimated. The reader can only guess that the study of the fossil record or the study of biological diversity is *somehow* important.

Mayr also presents these issues alongside the distinction between *why* and *how* questions (Mayr, 1961, p. 1502). Herein lies another problem: the distinction between *why* and *how* questions is unclear and from an analytic perspective unconvincing (Sect. 4.2). In addition to this, he speaks about the significance of adaptations, but also in a vague and inconclusive manner:

> [The evolutionary biologist] studies the forces that bring about changes in faunas and floras (as in part documented by paleontology), and he studies the steps by which have evolved the miraculous adaptations so characteristic of every aspect of the organic world. (Mayr, 1961, p. 1502)

Reading this, one has to admit that Francis's suspicion that ultimate explanations are in fact functional analyses is not entirely unfounded. In that essay Mayr made no effort to explain how homologous and non-homologous traits are told apart or how it is possible to know that the warbler's disposition to leave New Hampshire at the end of summer is an effect of natural selection (Sects. 4.3 and 4.4).

Such epistemic questions are also avoided in Mayr's (1993) reply to Francis. Even more disturbing perhaps is that Mayr attacks Francis for using the dictionary in order to find out what 'ultimate' is usually supposed to mean. Mayr gives much room to this reply explaining that he changed his mind about how to name ultimate causes, now preferring to call them 'evolutionary causes'. This reply is quite beside the point and gives the impression that the problem was how to choose words that sound good.

Nevertheless, the analysis in this section has shown that Francis's attack on Mayr does not succeed. Although Mayr offers a weak and misguided defence of his position, Francis misses a point when he doesn't acknowledge that evolutionary biologists have a methodology for distinguishing adaptive traits from traits that are a mere consequence of common ancestry. One can and should criticise Mayr for not having a persuasive account of ultimate causes. However, the position that there are no ultimate causes is not tenable.

4.7 Are Ultimate Causes Reducible to Proximate Causes?

André Ariew (2003) has made another argument against the ultimate-proximate distinction. He argues that Mayr's distinction between ultimate and proximate causes needs to be *replaced* with a distinction between dynamical and statistical explanations. According to Ariew, Mayr is mistaken when he believes that '"ultimate" refers to a dynamical process or force rather than an explanation of general patterns within an ensemble of processes' (Ariew, 2003, p. 564). How does Ariew arrive at this conclusion? There are two lines of argument: the first is that Mayr's program account of ultimate and proximate causes is dispensable, the second is that Mayr's notion of ultimate cause is reducible to the notion of proximate cause. I shall argue that neither the first nor the second line of argument succeeds.

The key question underlying Ariew's first line of argument is whether the program account is needed in order to recognise proximate causes. The answer given to that question is that

> Mayr's definition of proximate cause in terms of a genetic program is superfluous. Proximate causes can be identified without reference to an information metaphor. (Ariew, 2003, p. 555)

Ariew's argument is that the program account is dispensable from the epistemic point of view. Mayr would *not* have disagreed with this, since he maintained that one can recognise the causal impact of DNA without having knowledge of DNA. This is evident if one considers Mayr's discussion of Aristotle. But it is of course also true that Mayr's program account is in tension with the much more plausible temporal criterion that one might use as a means of distinguishing ultimate from proximate causes. I, therefore, agree with Ariew that one has to abandon the program account (although I am discarding it for a different reason).

Ariew also takes issue with Mayr's concept of proximate cause. Here Ariew's view is that 'proximate cause' is too narrow a concept. It focuses on the role of DNA, but it neglects the role of 'complex interactions between genes, extra-cellular mechanisms and environmental conditions' (Ariew, 2003, pp. 555 f.). The study of ontogenetic development, Ariew argues, is not just a matter of studying DNA. He argues that Mayr should adopt

> a broader view of development whereby proximate causes range over the various cellular and environmental conditions that constitute development as well as the actions of the DNA molecules. However, once we acknowledge that development can be understood as a *causal process* involving more than DNA molecules, the concept of a "genetic program" becomes superfluous. (Ariew, 2003, p. 556, italics in original)

According to this argument, the program account is no longer needed, since it would single out the wrong, i.e. too narrow, extension of 'proximate cause'. In other words, Ariew opts for a refined concept of proximate cause with a more inclusive extension. This reasoning is similar to Francis's reasoning that Mayr's concept of ultimate cause is too narrow compared to the concept of phylogenetic cause. In both cases the objection is the same: the concept has to be more inclusive.

Consequently, the reply can be the same: Why wouldn't it be enough to clarify extensional differences between different concepts? How should it be possible to force someone to choose between concepts instead of admitting that either concept is legitimate? The fact that a more inclusive concept is conceivable does not in itself constitute a valid objection to an alternative concept. Hence, demonstrating that a more inclusive concept than Mayr's notion of proximate cause is conceivable does not refute Mayr's program account.

Ariew's second line of argument is to show that Mayr's notion of ultimate cause reduces to the notion of proximate cause. The starting point of this argument is that (this time) the notion of ultimate cause has to be more inclusive, i.e. it should cover genetic drift and other factors such as migration, mutation and genetic recombination. This criticism is definitely justified and in accordance with my analysis in Sect. 4.4. However, Ariew would not be satisfied with minor corrections, such as I suggested above. His argument eventually culminates in a proposal to replace the concept of ultimate cause completely (Ariew, 2003, pp. 557 f.).

Ariew reasons that evolutionary theory does not deal with 'dynamical' processes. A dynamical process, as understood by Ariew, is a causal process taking place between individuals and their environment. Ariew, however, believes that evolutionary explanations are different: they 'range over statistical attributes of a population, not dynamical properties of individuals' (Ariew, 2003, p. 560).

Ariew thinks that this difference needs to be acknowledged when one tries to give evolutionary explanations. He argues that explanations in terms of proximate and ultimate causes are not distinct. This argument was impossible if one accepted Mayr's supposition that ultimate explanations are dynamical (i.e. causal) explanations, too. Ariew's upshot is that ultimate explanations are reducible to proximate explanations in the sense that a lengthy enumeration of proximate causes can replace an explanation in terms of ultimate causes. He suggests that, by upholding the view that evolutionary explanations are statistical explanations, one is in a position to maintain that evolutionary explanations and proximate (i.e. dynamical) explanations are distinct kinds of explanation (Ariew, 2003, pp. 560 ff.).

Ariew's reason for arguing that it would be unsatisfactory to reduce ultimate explanations to proximate explanations is that

> after tracing out proximate causes for every individual in a population, there is something left over to explain, namely what some of these disparate life histories have in common that set them apart from their conspecifics. (Ariew, 2003, p. 561)

Statistical explanations would explain what is not covered by those dynamical-reductive explanations. A statistical explanation would tell one how a certain trait affects the survival of individuals *on average*. This, however, meant that one was dealing with the properties of populations, not the properties of individuals. For instance, no single couple would exhibit the property of having 2.3 children. Yet in a population it could be true that those who possess a certain trait have on average 2.3 children. Ariew argues that evolutionary theory deals with such population-level and not individual-level properties (Ariew, 2003, p. 562).

The conclusion following from this line of argument is that evolutionary explanations are no causal explanations in the sense that they are not dynamical. This view, which is further elaborated in Matthen and Ariew (2009), differs radically from Mayr's original account of ultimate and proximate causes: if one accepted the suggestion of replacing explanations in terms of ultimate and proximate causes with explanations in terms of statistical relevance and dynamical (i.e. causal) interactions, one would give up the concept of ultimate causes.

However, I don't think that this critique is detrimental to Mayr's position. While I agree that there is a sense in which ultimate causes reduce to proximate causes, there is also a sense in which this need not be the case. However, in order to build such an argument one cannot take Mayr's original explanation of what ultimate causes are as a foundation. That account is too incoherent. What one might discuss is whether an account that commits itself unambiguously to Thomson's temporal criterion might be defended against the accusation that it reduces ultimate causes to proximate causes.

Ariew's position is non-reductionist for the simple reason that it distinguishes a non-causal (statistical) explanation from a causal (dynamical) explanation. I propose that Mayr's position (or rather a Thomson-like interpretation of it) is then also non-reductionist if ultimate and proximate cause are distinct concepts. How can an explanation in terms of ultimate causes reduce to an explanation in terms of proximate causes?

The extensional analysis in Sect. 4.5 shows that both proximate and ultimate causes can be seen as subsets of the broader ordinary notion of a dominant cause. That could mean that explanations in terms of both ultimate and proximate causes 'reduce' to explanations in terms of ordinary causation in the sense that they are merely special cases. That, however, would not yet show that ultimate causes reduce to proximate causes! In order to answer that question one has to know how the extensions of 'ultimate' and 'proximate cause' are related: are they coextensional, intersecting or possibly completely disjoint?

For some non-causal concepts such a decision can be straightforward. Take as an example the extensions of the concepts 'glucose' and 'water'. They are disjoint, since nothing in the extension of 'glucose' is also in the extension of 'water' and vice versa. The extension of 'glucose' is the set of all $C_6H_{12}O_6$ molecules, while the extension of water is the set of all H_2O molecules.

While 'glucose' or 'water' refer to intrinsic properties, causal concepts refer to relational properties. If one were to conceive of 'ultimate' and 'proximate cause' as intrinsic properties of certain events, this would rather obscure what the difference between those concepts is supposed to be. According to my above analysis (Sect. 4.5) one would have to compare the extensions of C^3_{51} and C^3_{53}. That those extensions are disjoint is easy to see, since C^3_{51} only takes the normal conditions f_1 and f_2, while C^3_{53} only takes the normal conditions f_3 and f_4.

One may argue, however, that this reasoning depends too heavily on the role of the normal conditions, which could be seen as problematic given that different normal conditions might (at least sometimes) still select the same cause-effect pairs. Then C^3_{51} and C^3_{53} would still be technically disjoint but not in any compelling or significant way. However, what should also be taken into account is that this problem of coextensional cause-effect pairs is a consequence of Mayr's inconsistent application of Thomson's temporal criterion, which he apparently accepted and at the same time inadvertently undermined.

If one upheld Thomson's criterion, the problem would not arise. In order to see this clearly one might stipulate extensions of 'ultimate' and 'proximate cause' without normal conditions

C^2_{50}: ① is an ultimate cause of ②
C^2_{52}: ① is a proximate cause of ②

and stipulate that ultimate causes occur *before* and proximate causes *during* the lifetime of an individual. It is then no longer possible that there is any relation in the extension of C^2_{50} that is also an element of the extension of C^2_{52}.

In order to achieve consistency one *must not* relapse into Mayr's confused thinking. Ultimate and proximate causes in this sense are solely defined in terms of their position in time. One *cannot* (but also need not) sort these cause-effect pairs into the categories of ecological, genetic, extrinsic physiological and intrinsic physiological cause.

Naturally, the warbler also takes part in events of future evolution. Yet those causes would all affect *future* generations. Hence nothing that happens to the warbler would be an ultimate cause of what happens to the warbler that exists now. This should suffice to show that there are no relations in the extension of 'ultimate cause' that are also in the extension of 'proximate cause', which means that a consequent application of Thomson's criterion results in two disjoint concepts of cause. Given such an interpretation of ultimate and proximate causes, 'ultimate cause' cannot be reduced to 'proximate cause' by conceptual subsumption. While the two concepts are subsumable to a broader ordinary notion of cause, they are not subsumable to one another.

Hence, there are two options for avoiding the conclusion that the extensions of ultimate and proximate cause intersect: a trivial solution relying on the distinctness of normal conditions and a less trivial suggestion that revives Thomson's temporal criterion at the expense of discarding much of Mayr's thoughts of how to put these causal relations into categories.

The way Ariew sees it, 'reduction' means that explanation in terms of ultimate causation is reducible to an explanation in terms of proximate causes in the sense that the one is replaceable with the other salva veritate. I agree that ultimate explanations are in this sense reducible to proximate explanations.

One might compare this reasoning to my argument in Sect. 3.3 where I argued that the concept of mixed cause can be considered as a means of syntactically simplifying a more complex disjunctive causal judgement. There is, however, one significant difference: 'mixed cause', which is shorthand for 'not being a positive or negative cause nor being causally neutral', replaces a *unique* syntactic structure, while 'ultimate cause' does *not* replace a unique syntactic structure.

It is important to see that this is *not* what happens when one adopts Thomson's criterion. Consider the paradigm example of bird migration again: Naturally, there is more than one event that is an ultimate cause of the warbler's departure from New Hampshire on the night of the 25th of August. The warbler has had thousands or millions of ancestors, depending on how many generations are taken into consideration. An event such as a mutation in germ-line cells, and the separation and recombination of such an altered allele with other alleles, might have taken place 100 generations ago, but also 99 generations ago or 101 generations ago, or 204,521 generations ago. There is no specific number of generations that elapses between an ultimate cause and its present-day effect. As a consequence, it is impossible to substitute an alternative expression for 'ultimate cause' in the same way that it can be done for 'mixed cause': there is a general solution for 'mixed cause' but no general solution for 'ultimate cause'.

Consider the following:

d_{25}: departure of the warbler on the night of the 25th of August
a_{12}/a_{22}: event in the past
$b_{13}/b_{14}/b_{15}/c_{23}/c_{24}$: intermediate event

Now assume that a_{12} is an ultimate cause of d_{25} and that a_{22} is an ultimate cause of d_{25}, too:

$$C^2_{50}a_{12}d_{25}\,, \tag{4.3}$$

$$C^2_{50}a_{22}d_{25}\,. \tag{4.4}$$

Ariew claims that any such statement is reducible in the sense that one can replace it with an explanation in terms of proximate causes. Assume that d_{25} is connected to its ultimate causes via two causal chains such that the following expressions in terms of proximate causation are true:

$$C^2_{52}a_{12}b_{13} \wedge C^2_{52}b_{13}b_{14} \wedge C^2_{52}b_{14}b_{15} \wedge C^2_{52}b_{15}d_{25}\,, \tag{4.5}$$

$$C^2_{52}a_{22}c_{23} \wedge C^2_{52}c_{23}c_{24} \wedge C^2_{52}c_{24}d_{25}\,. \tag{4.6}$$

The two statements formulated in terms of ultimate causation are reducible in the sense that the following expressions are true:

$$C^2_{50}a_{12}d_{25} \leftrightarrow C^2_{52}a_{12}b_{13} \wedge C^2_{52}b_{13}b_{14} \wedge C^2_{52}b_{14}b_{15} \wedge C^2_{52}b_{15}d_{25}\,, \tag{4.7}$$

$$C^2_{50}a_{22}d_{25} \leftrightarrow C^2_{52}a_{22}c_{23} \wedge C^2_{52}c_{23}c_{24} \wedge C^2_{52}c_{24}d_{25}\,. \tag{4.8}$$

In reality one would need much longer conjunctive statements. However, the point is that these conjunctive statements vary in length. It is only possible to reduce statements in terms of ultimate causes to statements in terms of proximate causes on a *case-by-case* basis.

What does this mean for Ariew's critique of ultimate causes? Since Mayr's account is unclear, it is also unclear how the charge of reductionism is applicable. All I can show here is that reduction, if it occurs, is likely to take place on a case-by-case basis. Under a strict Thomson interpretation, 'ultimate cause' doesn't reduce to 'proximate cause' in the same way as 'mixed cause' reduces to 'not being a positive or negative cause nor being causally neutral'. One can of course debate whether I offer too much help to Mayr by going back to Thomson. I see it, nevertheless, as a reasonable attempt at rescuing Mayr's account, since it historically developed from Thomson's thoughts.

Eventually, one is left with Ariew's objection that Mayr's account has a historical bias, i.e. too strong an inclination to examine past events instead of examining present and potential future events:

> Mayr is overly restrictive about the domain of evolutionary biology. He provides no reason to think that evolutionary biologists should not ask non-historical questions about, e.g. traits that are *currently* undergoing evolution. [...] Mayr is mistaken about evolutionary explanations being essentially historical. Rather, heritable differential reproductive success is an ongoing process. (Ariew, 2003, pp. 558 f., italics in original)

There is little here with which to disagree. Mayr is clearly more interested in the past than in the present. According to Mayr, evolutionary biology is a historical science. The extinction of the dinosaurs is an example of evolutionary history of which we have knowledge but which no one has witnessed. Although one may entertain hypotheses about the extinction of the dinosaurs, i.e. 'historical narratives' (Mayr, 2004, p. 32), one cannot test these hypotheses in a laboratory experiment. The only way left is to argue for or against a hypothesis on the basis of the evidence that still exists. It is only possible to conjecture which of the various hypotheses is true, e.g. the thesis that they fell victim to an epidemic, the thesis that they fell victim to a climatic catastrophe, or the thesis that they died in the aftermath of an asteroid impact (Mayr, 2004, pp. 32 f.).

Although it is true that Mayr has an inclination to deal with the history of life instead of present-day events, this is still no argument against Mayr. There is nothing in his account (not even in his confused version) that would force him to deny that present-day events are ultimate causes of future effects in evolution. What Ariew identifies as a 'historical bias' is perhaps nothing more than the observation that Mayr predominately discussed macroevolution, i.e. how major branches of the tree of life evolved, while Ariew himself has a greater interest in microevolution, i.e. how populations evolve on a shorter timescale. One can then argue that Ariew again invokes a false dichotomy—like the dichotomy between Mayr's notion of proximate cause and Ariew's refined notion of proximate cause—when he suggests that evolutionary explanations have to be either 'statistical' or 'dynamical'.

While Ariew states various valid points of criticism of Mayr's account, I would not yet accept Ariew's conclusion that one has to abandon the idea of ultimate and proximate causes altogether. Ariew's critique suffers from weaknesses similar to those in Francis's arguments. Both Francis and Ariew suggest that there is a choice between mutually exclusive options, when it is in fact possible for one to co-exist with the other.

4.8 Leaving the Ultimate-Proximate Account Behind

How to deal with the legacy of Mayr's account is still debated. A focal point in the discussion is the role it played during the so-called Modern Synthesis. (The expression 'Modern Synthesis' is derived from the title of a work by Julian Huxley

(1942): *Evolution. The Modern Synthesis.*) The synthesis was a process by which researchers from various biological sub-disciplines, who previously had rarely communicated with each other, recognised that they could combine the results of their research into a new, comprehensive theory of evolution (Mayr, 1982, p. 119). In Mayr's recollection, a conference at Princeton University in 1947 marked the culmination of that development (Mayr, 1980/1998, p. 42).

The theory that emerged from this process was the synthetic theory of evolution, 'that resulted from a combination of genetics, systematics, comparative morphology and palaeontology' (Müller, 2007, p. 946). However, a widely held view is that developmental biology (i.e. the study of ontogeny) wasn't a part of this process, and that a divide occurred between the synthetic theory on one side and developmental biology on the other. There have been various attempts to analyse why developmental biology was neglected in the synthesis and how evolutionary biology or developmental biology can be reconnected in the new field of so-called evo-devo biology (e.g. Love, 2003; Amundson, 2005; Müller, 2007; Carroll, 2008; Love, 2009; Pigliucci, 2009).

One hypothesis is that Mayr's ultimate-proximate distinction contributed substantially to the neglect of developmental biology in the synthesis. Laland et al. think that Mayr's account of ultimate and proximate causes 'has proven problematic because it builds on an incorrect view of development' (Laland et al., 2011, p. 1515). According to them, it is Mayr's mistake to view 'organisms as "programmed" by selection' (Laland et al., 2011, p. 1513). In their opinion it is far too simple to assume that new variants in a population arise only by random mutation (i.e. random changes in genetic programs). Ontogenetic development, they argue, can favour and disfavour the occurrence of new traits, thus changing the course of evolution by making some traits available to natural selection and others unavailable (Laland et al., 2011, p. 1513).

They also conduct a detailed discussion of co-evolution, i.e. the evolution of traits that change their environment, which then changes traits in return. A paradigm example they use is how sexual selection creates the peacock's tail. This is usually considered to be the evolutionary outcome of the mating preference of the peahen, which is in turn a subject of evolution and not an unchanging property of the environment (Laland et al., 2011, p. 1512). Another example is niche construction: earthworms are adapted to their soil environment, but ancestral earthworms are also a cause of that soil environment (Laland et al., 2011, p. 1514).

The point of these examples is to argue that Mayr cannot account for feedback in complex systems:

> Mayr's unidirectional characterization of causation encourages focus on single cause-effect relations within systems rather than on broader trends, feedback cycles, or the tracing of causal influences throughout systems (Laland et al., 2011, p. 1516).

Feedback is an important concept, yet it is also true that causation doesn't run in circles. That would require backward causation. In a technical sense, feedback cannot be a circle but only a succession of token causes instantiating a repeating pattern of types. Therefore, the concern that Mayr's account is 'unidirectional' cannot be meant in a token sense. It has to be about how type causes get arranged in models, i.e. how different types give input to each other. When Laland et al. object to 'single cause-effect relations', this must also apply to types, i.e. they insist that there are no exclusive relations between types.

Viewed that way, the reasoning is suddenly familiar. Mill also criticised Hume's naive regularity theory (Sect. 2.1) for the reason that it assumed exclusive regularities between types, rather than accepting the fact that type effects can have many type causes (Millian plurality of causes principle).

Laland et al. are putting forward an analogous thesis by arguing that traits are not exclusively caused by ultimate causes (in the type sense). They say about the peacock's tail that '[t]he cycle of causation may have begun with a prior preference or with fitness differences in a trait' (Laland et al., 2011, p. 1512). This, however, is not yet detrimental to Mayr's account. From Mayr's perspective one can easily acknowledge that proximate causes are causes of traits, too. Mayr can then still argue that proximate causes must in turn be traced back to ultimate causes. Dickins and Barton have for this reason objected that it is 'somewhat trivial' to argue about the starting point of explanations that run in circles (Dickins & Barton, 2013, p. 751).

The simplicity of this mental image of a cycle of type causes is indeed a problem, since it obscures how Laland et al. might succeed with their arguments against Mayr. They believe that there are various predictions of empirical fact on which adherents of the ultimate-proximate distinction and sceptics of that position differ (Laland et al., 2013, p. 801). How could that be possible within the limits of such a simple model?

An issue that may help illuminate the problem is the history of the prediction and discovery of homologous genes. Ron Amundson has provided an in-depth analysis of that dispute.[5] Amundson reports that Mayr had great faith in the almost unlimited power of natural selection. Mayr predicted therefore that nature would invent ever-newer genes as the molecular underpinning of phenotypes. A consequence of this view is that one cannot expect to find genes that are widely shared between species because of common ancestry, i.e. homologous genes (Amundson, 2005, pp. 213 ff.).

The present-day knowledge is, however, that such genes exist, i.e. that there are genes with almost the same DNA sequence because of common ancestry. Most famous is *pax-6*, a gene that controls eye-development in flies and mice (Stearns & Hoekstra, 2005, p. 147). Initially, genes could only be detected and traced by breeding experiments. This is only possible if there are at least two alleles (alternative forms of the same gene) that produce distinguishable phenotypes such as difference in colour or shape. Mendel, the inventor of this experimental approach, only examined those traits that he could *clearly* distinguish, setting aside those traits that blend into each other (Mendel, 1866, p. 5). These are problems when one is dealing with only one species. When it comes to homologous genes, i.e. genes that are shared between different species as a consequence of common ancestry, the

[5] Laland et al. (2013, pp. 796 f.) partially rely on Amundson's results, which makes his analysis an apt topic for discussion.

problem is even worse: because of intersterility it is impossible to carry out breeding experiments (Amundson, 2005, p. 215).

Mayr was of course familiar with the inherent limitations of the Mendelian approach, i.e. the Mendelian blind spots that have only been removed as sequencing DNA has become an option. However, Mayr did not believe that it was likely that homologous genes would exist. Amundson cites this as an example that 'illustrates an aspect of scientific commitment that extends well beyond the data immediately at hand' (Amundson, 2005, p. 217). Mayr's prediction that there are no homologous genes doesn't follow directly from his distinction between ultimate and proximate causes, yet Mayr may have believed 'that the irrelevance of development to evolution follows directly from the distinction itself' (Amundson, 2005, p. 223).

But even if one assumes that Mayr did not believe that the existence of homologous genes would be a simple matter of conceptual implication, it is easy to see how the denial of homologous genes and the ultimate-proximate distinction fit together. If you uphold the doctrine that there are 'two largely separate fields which differ greatly in method, *Fragestellung*, and basic concepts' (Mayr, 1961, p. 1501), you exempt yourself from considering the possibility that a careful examination of molecular relationships (proximate causes) could unearth something so intriguing as homologous genes.[6]

Amundson believes that the ultimate-proximate distinction in itself does not rule out a third category of ontogenetic causes. He suggests that one could 'conceive of ontogeny as an in-between point in the proximate-ultimate scale' (Amundson, 2005, p. 204). Speaking of the ultimate-proximate distinction, Amundson points out that

> [i]t illustrates (but does not name) the principle of explanatory relativity. When presuppositions differ, it is possible that (what sounds like) the *very same fact* is explained by apparently inconsistent explanations [...]. The contrast between proximate and ultimate causation is a careful and specific recognition of the relativity of explanation. [...] That's the good side of the proximate-ultimate distinction. (Amundson, 2005, pp. 203 f., italics in original)

Although Amundson might think of explanatory relativity as a benefit and a third category of causes as an improvement, I don't agree. If a circle of ultimate and proximate causes is trivial, a circle of ultimate, ontogenetic and proximate causes is also trivial. Even worse, if explanatory relativity is a virtue, this strengthens Dickins and Barton's position that choosing the starting point is *trivial*.

I would even go so far as to claim that explanatory relativity is the venom in Mayr's account. Amundson observes that Mayr used the ultimate-proximate distinction 'after 1970 to categorize developmental processes as proximate, and so to label them as logically irrelevant to ultimate evolutionary explanations' (Amundson, 2005, p. 212). How can explanatory relativity be a good thing if Mayr weaponised it in the dispute over developmental biology? Shouldn't that tell us that there is something wrong?

---

[6] In this context, 'Fragestellung' translates to 'research question'. There doesn't appear to be any deeper meaning intended.

The problem, as I see it, is that explanatory relativity gets conflated or mixed up with context sensitivity. The notion of a dominant cause (Sect. 2.6) is context sensitive, but it is *not* relative in the sense that anything can be a cause. Why was it such a surprise finding homologous genes shared between only distantly related species? It is only astonishing if you assume it to be *normal* that proximate mechanisms provide a vast amount of non-homologous genes on which selection can act.

Had it been as Mayr predicted, the statistical majority of species would *not* have shared genes with other species, i.e. variation in non-homologous genes would have been normal. In this case phenotypic appearances would be largely determined by selection. Homoplasies are common in this scenario.

As has been found, however, the statistical majority of species does share genes with other species, i.e. variation in non-homologous genes isn't normal. In this case phenotypic appearances are to a greater extent, although of course not exclusively, owed to a lack of variation in non-homologous genes. Homoplasies are less common under these circumstances.

I am not claiming that a theory such as Hart and Honoré's could be straightforwardly applied here. A more nuanced account is likely to be needed in order to weigh contributions instead of naming one dominant cause. However, distinguishing between causes and *conditions* (even if done in degrees) is vital to any compelling argument trying to show why Mayr made a mistake in pushing ontogenetic development aside.

Amundson seems to suppose that Mayr could have avoided the misuse of the ultimate-proximate distinction quite simply, had he committed himself thoroughly to principles of explanatory relativity. The problem with this idea is that explanatory relativity has a limitation and a disadvantage. It might work as a *weak* defence if one is fighting a losing battle, as Mayr did against molecular biology. As mentioned earlier, John Beatty reports that Mayr used this strategy on many occasions, including as means of securing his funding when he was Director of Harvard's Museum of Comparative Zoology (Beatty, 1994, pp. 351 f.). This might have delayed but could not prevent the triumph of molecular biology.

The disadvantage of explanatory relativity is that it is an *even weaker* offensive argumentative strategy. Explanatory relativity (but not context sensitivity) gives equal weight to all perspectives. I could for example propose that it would be an explanation of someone's death to say that there was a particular atmospheric condition that caused that death (Sect. 2.1). If it is true that the death in question would not have occurred without that atmospheric condition being present, I have produced a true, although uninteresting, explanation. If I wanted to convince others of the value of my explanation by pointing to the atmospheric condition as a cause of a particular death, I could say, 'Look, here is the principle of explanatory relativity, so would you please consider my explanation.' My opponent, however, could *reasonably* reply, 'Thanks, but I am not interested.' Likewise, if the scientist who is interested in ontogenetic causes had said, 'Look, this is the principle of explanatory relativity, so would you please consider my explanation,' Mayr could have *reasonably* replied, 'Thanks, but I am not interested.'

Explanatory relativity is a truism few would want to dispute. I, too, would not categorically reject the idea that explanatory relativity might be a valid stance under different circumstances. If a phenomenon is explained to a number of people who each have different background knowledge, this can justify tailoring explanations to their specific perspectives. This should not be confused with the common assumption that, all explanation being interest driven, any one explanation is as good as any other.

The question of what shapes the phenotype can be answered differently, but not all answers are based on objective facts. It is not a valid reply to argue that Mayr's position would still be true if one took a different perspective. What Mayr got wrong is that he assumed that proximate causes are mere proxies without any potential of influencing the outcome of what ultimate causes dictate to them. Proximate causes do not allow variation to occur to the degree presumed by Mayr. Note that there is still a trivial sense in which Mayr's thesis remains true: even from today's perspective any ultimate cause qualifies as a condition. However, that is *not* the relevant notion when biologists pose the question of how much the phenotype is influenced by selection, development or—also not to be forgotten—environmental factors.

Once this is clarified, one can quite easily explain how it is possible for so many to have accepted Mayr's distinction between ultimate and proximate causes without question: Mayr first alludes to a broad notion of cause where there are no differences between causes and conditions; he then divides this set of causes into 'ultimate' and 'proximate causes'; and finally he claims—without any argument—that the ultimate category contains the causes and the proximate category contains the conditions. One should of course not assume that this was a consciously devised evil master plan. It is much more likely that this is an instance of unintentionally misapplying the idea of explanatory relativity in an unreasonable way. After all, he did not discriminate between causes and conditions, calling everything a 'cause'.

4.9 Conclusion

There are many inconsistencies in Mayr's account of ultimate and proximate causes, which is what makes an assessment so difficult. If Mayr's account weren't so influential, it could easily be dismissed for the sole reason that it is remarkably incoherent. Since it has had so many followers, however, one cannot leave it behind without an analysis of what made it so successful. This is what I have tried to do in this chapter by first looking at its origins, then reconstructing a plausible version of it, and finally also defending it against criticism. In the end, however, I too have to conclude that the account of ultimate and proximate causes is hopelessly flawed. Once you dismantle explanatory relativity as the only—presumably—positive aspect of Mayr's account, nothing is left. I therefore applaud anyone who definitively abandons the Mayr account of ultimate and proximate causes, once and for all.

References

Amundson, R. (2005). *The changing role of the embryo in evolutionary thought. Roots of evo-devo*. New York: Cambridge University Press.

Ariew, A. (2003). Ernst Mayr's 'ultimate/proximate' distinction reconsidered and reconstructed. *Biology & Philosophy, 18*(4), 553–565. https://doi.org/10.1023/A:1025565119032

Baker, J. R. (1938). The evolution of breeding seasons. In G. R. de Beer (Ed.), *Evolution. Essays on aspects of evolutionary biology* (pp. 161–177). Oxford: Clarendon Press.

Beatty, J. (1994). The proximate/ultimate distinction in the multiple careers of Ernst Mayr. *Biology & Philosophy, 9*(3), 333–356. https://doi.org/10.1007/BF00857940

Calcott, B. (2013). Why how and why aren't enough: More problems with Mayr's proximate-ultimate distinction. *Biology & Philosophy, 28*(5), 767–780. https://doi.org/10.1007/s10539-013-9367-1

Carroll, S. B. (2008). Evo-devo and an expanding evolutionary synthesis: A genetic theory of morphological evolution. *Cell, 134*, 25–36. https://doi.org/10.1016/j.cell.2008.06.030

Cummins, R. (1975). Functional analysis. *The Journal of Philosophy, 72*(20), 741–765. https://doi.org/10.2307/2024640

Cummins, R., & Roth, M. (2010). Traits have not evolved to function the way they do because of a past advantage. In F. J. Ayala & R. Arp (Eds.), *Contemporary debates in philosophy of biology* (pp. 72–85). Chichester, West Sussex: Wiley-Blackwell.

Dickins, T., & Barton, R. (2013). Reciprocal causation and the proximate–ultimate distinction. *Biology & Philosophy, 28*(5), 747–756. https://doi.org/10.1007/s10539-012-9345-z

Driesch, H. (1908). *The science and philosophy of the organism*. Aberdeen: Aberdeen University.

Francis, R. C. (1990). Causes, proximate and ultimate. *Biology & Philosophy, 5*(4), 401–415. https://doi.org/10.1007/BF02207379

Haig, D. (2013). Proximate and ultimate causes: How come? and what for? *Biology & Philosophy, 28*(5), 781–786. https://doi.org/10.1007/s10539-013-9369-z

Huxley, J. (1942). *Evolution. The modern synthesis*. London: George Allen & Unwin Ltd.

Laland, K. N., Odling-Smee, J., Hoppitt, W., & Uller, T. (2013). More on how and why: A response to commentaries. *Biology & Philosophy, 28*(5), 793–810. https://doi.org/10.1007/s10539-013-9380-4

Laland, K. N., Sterelny, K., Odling-Smee, J., Hoppitt, W., & Uller, T. (2011). Cause and effect in biology revisited: Is Mayr's proximate-ultimate dichotomy still useful? *Science, 334*(6062), 1512–1516. https://doi.org/10.1126/science.1210879

Love, A. C. (2003). Evolutionary morphology, innovation, and the synthesis of evolutionary and developmental biology. *Biology & Philosophy, 18*(2), 309–345. https://doi.org/10.1023/A:1023940220348

Love, A. C. (2009). Marine invertebrates, model organisms, and the modern synthesis: Epistemic values, evo-devo, and exclusion. *Theory in Biosciences, 128*(1), 19–42. https://doi.org/10.1007/s12064-009-0063-2

Matthen, M., & Ariew, A. (2009). Selection and causation. *Philosophy of Science, 76*(2), 201–224. https://doi.org/10.1086/648102

Mayr, E. (1961). Cause and effect in biology: Kinds of causes, predictability, and teleology are viewed by a practicing biologist. *Science, 134*, 1501–1506. https://doi.org/10.1126/science.134.3489.1501

Mayr, E. (1974). Teleological and teleonomic, a new analysis. In R. S. Cohen & M. W. Wartofsky (Eds.), *Methodological and historical essays in the natural and social sciences* (Vol. XIV, pp. 91–117). Dordrecht: D. Reidel Publishing Company.

Mayr, E. (1982). *The growth of biological thought. Diversity, evolution, and inheritance*. Cambridge, MA and London, England: Belknap Press.

Mayr, E. (1993). Proximate and ultimate causations. *Biology & Philosophy, 8*(1), 93–94. https://doi.org/10.1007/BF00868508

Mayr, E. (1980/1998). Prologue: Some thoughts on the history of the evolutionary synthesis. In E. Mayr & W. B. Provine (Eds.), *The evolutionary synthesis. Perspectives on the unification of biology* (pp. 1–48). Cambridge, MA and London, England: Harvard University Press. (Original work published 1980)

Mayr, E. (2004). *What makes biology unique? Considerations on the autonomy of a scientific discipline*. New York: Cambridge University Press.

Mayr, E., & Meise, W. (1930). Theoretisches zur Geschichte des Vogelzuges. *Der Vogelzug. Berichte über Vogelzugsforschung und Vogelberingung*, *1*(4), 149–172

Mendel, G. (1866). Versuche über Pflanzen-Hybriden. *Verhandlungen des Naturforschenden Vereins zu Brünn*, *4*, 3–47.

Müller, G. B. (2007). Evo-devo: Extending the evolutionary synthesis. *Nature Reviews Genetics*, *8*(12), 943–949. https://doi.org/10.1038/nrg2219

Perlman, M. (2010). Traits have evolved to function the way they do because of a past advantage. In F. J. Ayala & R. Arp (Eds.), *Contemporary debates in philosophy of biology* (pp. 53–71). Chichester, West Sussex: Wiley-Blackwell.

Pigliucci, M. (2009). An extended synthesis for evolutionary biology. *Annals of the New York Academy of Sciences*, *1168*(1), 218–228. https://doi.org/10.1111/j.1749-6632.2009.04578.x

Pittendrigh, C. S. (1958/1967). Adaptation, natural selection, and behavior. In A. Roe & G. G. Simpson (Eds.), *Behavior and evolution* (4th ed., pp. 390–416). New Haven and London: Yale University Press. (Original work published 1958)

Stearns, S. C., & Hoekstra, R. F. (2005). *Evolution. An introduction* (2nd ed.). Oxford, New York: Oxford University Press.

Thomson, A. L. (1924). Photoperiodism in bird migration. *The Auk*, *41*(4), 639–641. https://doi.org/10.2307/4074295

Thomson, A. L. (1926). *Problems of bird-migration*. London: H. F. & G. Witherby.

Watson, J. D., & Crick, F. H. C. (1953). A structure for deoxyribose nucleic acid. *Nature*, *171*(4356), 737–738. https://doi.org/10.1038/171737a0

Watt, W. B. (2013). Causal mechanisms of evolution and the capacity for niche construction. *Biology & Philosophy*, *28*(5), 757–766. https://doi.org/10.1007/s10539-012-9353-z

Chapter 5
Permissive and Instructive Causes

Abstract This chapter provides a semantic analysis of 'permissive' and 'instructive cause'. Based on that analysis, I examine James Woodward's claim that the biological permissive-instructive scale translates into causal specificity as influence. My conclusion is that Woodward is correct in conjecturing this. This means that the permissive-instructive scale cannot be an instance of pluralism. However, I also discuss why Woodward didn't succeed in illustrating his account properly. The problem is a superficial analysis of protein biosynthesis. I argue that a superficial analysis of that mechanism is also responsible for the dispute between Brett Calcott and Pierrick Bourrat over Woodward's proposal. In my opinion neither Calcott's nor Bourrat's criticism of Woodward is successful. I close this chapter with a discussion of how my analysis of the permissive-instructive terminology fits together with Woodward's analysis and Conrad Hal Waddington's model of the epigenetic landscape.

Keywords Causal influence · Causal specificity · Counterfactual theory of causation · Embryonic induction · Epigenetic landscape · Instructive cause · Interventionist theory of causation · Mechanism · Mutual causal information · Permissive cause · Protein biosynthesis · Waddington box

The distinction between permissive and instructive causes is usually traced back to Howard Holtzer's contribution at the 18th Hahnemann Symposium *Induction of Chondrogenesis: A Concept in Quest of Mechanisms*. In contrast to the (at least in theory) symmetric distinction between ultimate and proximate causes, there is a clear asymmetry between permissive and instructive causes: while permissive causes merely allow a response, instructive causes shape a response in detail.

The question of how to interpret this terminology has recently been the subject of a (quite heated) debate between Calcott (2017; 2019) and Bourrat (2019a; 2019b; 2019c). The current debate started when Calcott reacted to an earlier suggestion by Woodward (2010) that the distinction between instructive and permissive causes might be analysed in terms of causal specificity as defined in interventionist theory. I shall argue that Woodward was *right* to state that hypothesis, although he used a

K. Ehrenstein, *Causal Pluralism in the Life Sciences*,
History, Philosophy and Theory of the Life Sciences 25,
https://doi.org/10.1007/978-3-030-87942-6_5

clumsy analogy as illustration. The confusion that followed is in my opinion due to a superficial philosophical analysis of protein biosynthesis and a poor analysis of how biologists have used the terminology.

In order to cut through this thicket, I shall first go back to the sources referenced by the participants in the debate (Sect. 5.1). A major problem is that a twofold shift occurred: on the one hand, there are authors who reinterpret the terminology—presumably because the original usage has become useless for them; on the other hand are authors who don't need to reinterpret the terminology since they are no longer examining the particular problem that gave rise to the distinction of permissive and instructive causes. Although I acknowledge that there are different applications of the terminology, I shall take Holtzer's position as a reference point for my arguments, since he articulates most clearly what 'permissive' and 'instructive' are supposed to mean.

5.1 Versions of the Permissive-Instructive Distinction

Many of the arguments exchanged by Bourrat and Calcott hinge on the semantics of 'permissive' and 'instructive'. Speaking of permissive causes, Calcott notes at one point 'that biologists have used the term loosely' (Calcott, 2019, p. 7). Yet neither Calcott nor Bourrat clearly analyses which alternative interpretations of 'permissive' and 'instructive' can be found in the sources they reference.

The initial problem raised by Holtzer—as may be deduced from his essay's title—is that there used to be a concept ('induction'), and it was unclear to which sort of mechanism it referred. In more philosophical or specifically Fregeian terms: there was a concept with a sense but unknown denotation.

Holtzer argued against a customary view of induction, previously also held by himself, which he describes as the position that undifferentiated, naive cells receive a signal from the inducing tissue in the form of a macromolecule, e.g. proteins or nucleic acids, which *instructs* the naive cells to differentiate, e.g. into chondrocytes. The exact chemical structure of the macromolecule in question may have been left open in such hypotheses. The notion of 'instruction' had been taken literally in the sense that it had been assumed that the macromolecule would carry information that specified in great detail how the naive cells would respond (Holtzer, 1968, p. 153 f.).

As a reaction to (then) new empirical data, Holtzer suggested that the inducing molecule might not carry as much information as previously assumed. He proposed that what looked like naive cells were actually 'covertly differentiated cells' that merely carried out responses for which they were *already* instructed. Another familiar way of expressing this is to say that the cells that responded were already competent (Holtzer, 1968, p. 154). Holtzer also notes that induction can occur as a response to more than one substance, which makes it unlikely that these substances specify the response in detail. This meant that the role of the inducing molecule was greatly diminished:

> In all these instances the exogenous agents do not transmit information in the sense that viral or pneumococcal nucleic acids transmit information. It is difficult to talk of the "informational" content of D_2O or ions. Rather the exogenous agents appear to release, to cue, preprogrammed, integrated sets of synthetic processes [. . .]. These extracellular agents *permit* cells to express activities determined by other events. (Holtzer, 1968, p. 161, italics added)

This outline of the instructive-permissive distinction and the reassessment of the causal role of inducing substances supports the historical analysis given by Marc W. Kirschner and John C. Gerhart, who point out (without referencing Holtzer in particular) that '[t]he search for inducers collapsed in confusion' once it turned out that inducers did not carry 'specific chemical information' (Kirschner & Gerhart, 2005, pp. 124 f.). The great surprise had been that 'embryonic induction turned out to be a permissive process' (Kirschner & Gerhart, 2005, p. 126).

The distinction between instructive and permissive causes, as Kirschner and Gerhart conceive it, is 'a matter of degree': a cause might act in a way that is somewhere midway along the instructive-permissive spectrum (Kirschner & Gerhart, 2005, p. 125). However, as far as embryonic induction is concerned, the attempts to find an inducing substance that is more on the instructive than the permissive side of the spectrum had all been unsuccessful (Kirschner & Gerhart, 2005, p. 127). The upshot is then that the distinction comes down to this: a stimulus that carries little specific information is a permissive cause, and a stimulus that carries a huge amount of detailed information is an instructive cause.

The terminology of instructive and permissive causes or interactions is used to this day. In the light of Kirschner and Gerhart's analysis that instructive causes in Holtzer's sense were never found, one must of course ask how this is possible. One may expect that the contemporary usage is perhaps not the same as Holtzer's original application.

In the most recent edition of their textbook *Developmental Biology* Michael J. Barresi and Scott F. Gilbert reference Howard Holtzer's article (1968) and explain the difference between instructive and permissive interaction, but not in the same way as Holtzer did. Their account is not given in terms of informational content but in terms of necessitation, which in my opinion doesn't help to clarify the distinction. According to them, 'instructive interaction' means that 'a signal from the inducing cell is *necessary* for initiating new gene expression in the responding cell' (Barresi & Gilbert, 2020, p. 99, italics in original). Permissive interaction, in contrast, is described as the case in which 'the responding tissue has already been specified and needs only an environment that allows the expression of these traits' (Barresi & Gilbert, 2020, p. 99). Both instructive and permissive causes are explained in terms of necessitation: in the instructive case there is a necessary signal involved, while in the permissive case there is a necessary environment involved.

More can be learned by taking Barresi and Gilbert's examples into account. They point to the induction of lens formation in the frog *Xenopus laevis* as a case of instructive interaction (Barresi & Gilbert, 2020, p. 99). This induction occurs when optic vesicles are placed underneath the head ectoderm. The cells of the ectoderm respond to paracrine factors (i.e. chemicals released by adjacent cells for communication, in this case the cells that form the optic vesicles), which results in the formation of the lens tissue.[1] Again I have to emphasise that Barresi and Gilbert are *not* talking about a transfer of information, but about what is necessary for this process to take place. This includes the prerequisite that the cells of the head ectoderm have to be competent. Placing optic vesicles underneath other parts of the ectoderm, such as the trunk, will not induce lens formation (Barresi & Gilbert, 2020, p. 96). Permissive interaction is illustrated with an experiment in which researchers succeeded in creating a beating heart by infusing the extracellular matrix of a decellularised (rat) heart with cardiomyocyte progenitor cells.[2] Barresi and Gilbert argue that in this case the interaction was permissive, since 'the environmental conditions of the decellularized extracellular matrix were equipped with *instructive* guidance for the development of heart muscle' (Barresi & Gilbert, 2020, p. 99, italics added). Note that permissive interaction is *also* explained in terms of instruction. Apparently, the difference between instructive and permissive interaction is *not* that the former involves instruction and the latter doesn't.

Bourrat, who refers to an earlier edition of Barresi and Gilbert's textbook, suggests that the role of permissive and instructive causes should be compared to the role of background conditions and triggering causes (Bourrat, 2019b, p. 3). I agree that this can be a valid interpretation of this textbook account of permissive and instructive causes. At least a distinction between different sources of signals is made by Barresi and Gilbert.

Given that Barresi and Gilbert diverge so much from Holtzer's account, one may want to know how the other sources referenced by Bourrat (2019b) besides Holtzer (1968) fit into this picture. Do they conform to Barresi and Gilbert's view or do they interpret the distinction in yet another way?

Surprisingly, the idea that instructive causes have to provide information, as envisaged by Holtzer, is articulated by Lemmon et al. (1992). Bourrat (2019b, p. 2) quotes them, citing a passage in which they are not explicitly speaking of information. I, however, believe that this passage from their abstract is more illuminating:

> These results suggest that some adhesion molecules may serve as permissive substrates in that they can define axonal pathways but they do not provide information about which path to take at a choice point or about which direction to go along the path. (Lemmon et al., 1992, p. 818)

Here the adhesion molecules are permissive causes in the sense that they are non-sufficient for directing axons in a particular direction, i.e. more information is needed for guiding them in a particular direction. This is an example that also hints at how one can speak of necessitation in this context. Although they don't speak explicitly of what is 'necessary' or 'sufficient', it is clear that their notion of

[1] Barresi and Gilbert reference (Grainger, 1992; Saha et al., 1989).

[2] Barresi and Gilbert reference (Ott et al., 2008).

a permissive cause is that of a non-sufficient but in some sense necessary condition. This is also what appears to be the gist of their discussion in the closing section, where they reiterate this idea and emphasise that adhesive molecules still have an important role: they might ‘promote rapid axon growth’ without directing axons in a particular direction (Lemmon et al., 1992, p. 825). This is clearly how one would view a permission from a common-sense perspective: something that is needed but which does not guarantee success. What an instructive cause is meant to be is not explicitly articulated. I assume, however, that an instructive cause has to be in some sense a sufficient cause, given some implicit background assumption, such as the undisturbed working of an organism.

How Hunter and Hatten (1995) apply the terminology is not entirely clear to me. In the passage quoted by Bourrat (2019b, p. 2) they speak of an instructive cause as the signal that commits a cell lineage to a specific cell type, while the permissive cause narrows down the cell fate to a certain range but not a specific option. I haven’t been able to deduce more from the rest of Hunter and Hutten’s article, but it appears to be at least consistent with Lemmon et al. (1992).

Another reference clearly not following Barresi and Gilbert is Crair (1999). The interesting fact about Crair is that he is discussing neither adhesion molecules nor the role of other chemicals released by cells as a means of cell–cell communication. This author is interested in the role of neural activity, i.e. signals forwarded by neurones. Although Crair applies the permissive-instructive terminology to a different kind of signal than that originally discussed by Holtzer, Crair’s usage of ‘permissive’ and ‘instructive’ matches Holtzer’s account perfectly.

The criterion used by Crair for distinguishing between permissive and instructive roles of neural activity involves asking whether a change in ‘the pattern or information content of the neural activity, while keeping the overall activity constant’ will result in different outcomes (Crair, 1999, p. 88). One experiment described is the keeping of amphibian embryos under a strobe in order to discover whether the neuronal activity resulting from strobe lights affects the outcome of their development (Crair, 1999, p. 90). This is virtually the same idea Holtzer had, i.e. the idea that a message has to be internally complex, and that this complexity must affect the outcome of the response, if it is to qualify as instructive. Crair also uses ‘permissive’ and ‘instructive’ as terms referring to opposite poles of a spectrum:

> The diversity of apparent effects of neural activity on brain circuit formation may simply define a continuum. This ranges from one extreme, at which neural activity is not at all necessary for the establishment of correct neural connections, passes through a permissive role for activity, and then, at the other extreme, the pattern of neural activity instructs the development of neural circuits. (Crair, 1999, p. 91)

This lends support to Kirschner and Gerhart’s above-mentioned observation that the terminology allows for gradual distinctions.

What is astonishing is how many applications of the permissive-instructive terminology there are, in the light of Kirschner and Gerhart’s indication that instructive causes were never found. Considering the examples discussed here, there appear to be two reasons why the terminology is still in use: one is that the terminology has been reinterpreted; the other is that the research topics have changed.

Authors such as Barresi and Gilbert created their own interpretation of what 'permissive' and 'instructive' mean. This is why they continue calling inducing substances 'instructive', despite the fact that the mechanism they are describing is clearly not instructive in Holtzer's sense. The inducers talked about in *Developmental Biology* are the sort of inducers postulated by Holtzer, i.e. they only release preprogrammed processes by virtue of fitting into receptors possessed by competent cells. Rather than what one may have conjectured from Holtzer's observation that induction can also be triggered by simple molecules, those inducers are quite complex, since they are proteins. Nevertheless, they don't induce because they transfer information in the sense of a construction plan, they induce because they bind to their respective and unique receptors initiating a signal chain that changes gene expression. Holtzer's original terminology is useless in this situation, since it classifies all these inducers in the same category, i.e. permissive. Other authors such as Crair need not reinterpret Holtzer's terminology, since they examine phenomena that can still be sorted by the original account.

5.2 The Notion of Influence Doesn't Apply to the RNA Polymerase

Woodward (2010) suggested in an essay on causation in biology that the notion of specificity, as understood in interventionist theory, may correspond to the distinction biologists make between permissive and instructive causes. This was, however, a mere suggestion to be explored further, and Woodward didn't provide any analysis of how biologists use these terms, only referring to the distinction as 'elusive' (Woodward, 2010, p. 317). That it is difficult to get a grasp on that terminology can only be agreed with (Sect. 5.1).

The focus of Woodward's analysis is how one can account for the different causal roles of the DNA sequence and the RNA polymerase in protein biosynthesis. Woodward's idea is that one might create a theoretical tool which could be part of an explanation for the alleged switch-like operation of the RNA polymerase, as opposed to the more specific mode of operation at work relating the DNA sequence to the mRNA transcript.

The tool offered by Woodward is an account of causal specificity. This idea is introduced by him as a 'variant' of Lewis's (2000/2004) notion of influence (Woodward, 2010, p. 301 f.). Lewis revised his original theory of counterfactual causation (as described in Sect. 2.2) by arguing that counterexamples to his theory could be refuted if one were to examine patterns of counterfactual dependence between sets of events instead of counterfactual dependence between two events.

Lewis stipulated that there is influence between distinct actual events C and E

> iff there is a substantial range C_1, C_2, ...of different not-too-distant alterations of C (including the actual alteration of C) and there is a range E_1, E_2, ...of alterations of E, at least some of which differ, such that if C_1 had occurred, E_1 would have occurred, and if C_2 had occurred, E_2 would have occurred, and so on. (Lewis, 2000/2004, p. 91)

As in the original theory causation is conceived of as an ancestral relation: that C causes E means that 'there is a chain of stepwise influence from C to E' (Lewis, 2000/2004, p. 91). Lewis believes that causal processes, as described by Reichenbach (Sect. 2.3) or Salmon (Sect. 2.5), are examples of special cases of patterns of causal influence (Lewis, 2000/2004, p. 91). He suggests that there are degrees of influence (in more than one dimension), i.e. one could ask how many alternatives C_i there are and how distant they are from C in the actual world (Lewis, 2000/2004, p. 92).

In contrast to Woodward, Lewis was not interested in the different causal roles of DNA sequence and RNA polymerase. One problem with Lewis's original theory was the counterexamples that resulted from late preemption. In a case of late preemption, as opposed to early preemption which was mentioned in Sect. 2.2, it is implausible to postulate that there is a causal chain of events that could be interfered with. One such example, given by Hall (2004, p. 235), is that Billy and Suzy are playing a game of throwing stones at a bottle. Suzy throws and her stone shatters the bottle. Billy also throws almost simultaneously, but when the stone arrives the bottle is already destroyed. In the original analysis this has the consequence that the effect (shattered bottle) dosen't counterfactually depend on Suzy's throw (the intuitively effective cause). This wouldn't be a problem if there were stepwise counterfactual dependence. Then the causal chain beginning with Suzy's throw could interfere with the causal chain beginning with Billy's throw before Billy's stone reaches the place where the bottle gets shattered. If Billy's causal chain doesn't run to completion, it cannot make a difference to counterfactual dependence. In this scenario, however, the causal chain beginning with Billy's throw cannot be interfered with and runs to completion. It is only preempted by the occurrence of the effect (shattered bottle).

With the notion of influence at hand Lewis can solve this problem by arguing that one can counterfactually fix either Billy's or Suzy's throw. If one holds fixed Billy's throw, Suzy's stone will still shatter the bottle and it will do so (in the nearby possible worlds) in a variety of different ways, i.e. Suzy aims a bit differently so that different fragments of glass are created. If, in contrast, Suzy's throw is fixed, the bottle will still shatter in exactly the same way as in the actual world. In the nearby possible worlds Billy's throw will differ in a variety of different ways, e.g. aiming a little bit higher or lower, but this has no consequences for the way the bottle shatters, which is always the same as in the actual world. Hence Suzy's throw has more influence and is therefore the cause of the shattered bottle (Lewis, 2000/2004, p. 92).

One can try to apply Lewis's revised counterfactual theory of causation to the problem of the causal roles of the DNA sequence and the RNA polymerase. The intuitive understanding is that the DNA sequence should have more influence. In order to see this, one holds fixed the event that the RNA polymerase is active and varies the event that there is a certain DNA sequence in nearby possible worlds. This is where one already encounters a problem: how should one 'fix' the activity of the RNA polymerase? In the Billy and Suzy example one can simply stipulate that Billy throws exactly the same way as in the actual world. If one does the same with the RNA polymerase this means that the RNA polymerase strings together exactly the same nucleotides as it does in the actual world regardless of the DNA template it is reading. This leads to the conclusion that the DNA sequence has *no influence* on the mRNA transcript. One could of course avoid that conclusion by demanding that it is not the particular activity token to be held fixed, but only the general activity type in which the RNA polymerase engages.

This latter suggestion would be in line with Lewis's preparedness to adjust standards of fragility. It could be that my initial suggestion is too harsh, since I treat the event as extremely fragile. One also has to take into account that even Billy's throw and Suzy's throw are not completely isolated from each other. Lewis points out that one has to assume that there are (very small) gravitational effects between the two events (Lewis, 2000/2004, p. 89). Not mentioning such details is due to—as Lewis sees it—harmless 'linguistic laxity' (Lewis, 2000/2004, p. 90).

However, that there might be some leeway for the adjustment of fragility must not obscure a crucial disanalogy between Suzy and Billy throwing stones at a bottle and DNA and RNA polymerase producing an mRNA transcript, which is that the former scenario is competitive, but the latter is cooperative.[3] If Suzy misses the target, Billy hits and destroys the bottle. He is not dependent on her and can do the work alone. Protein biosynthesis is different. RNA polymerase and DNA sequence are interdependent. One cannot do the work of the other without that partner. The events that one would want to compare with each other overlap to a great extent and cannot be altered independently. From this perspective, it starts to look as if it could be a good idea to adjust the idea of influence.

Woodward's solution within his interventionist approach (Sect. 2.4) is to have two notions of influence, of which the first is specificity as influence and the second one–one specificity. (I will return to the second notion in Sect. 5.4.) Both notions are used as a means of characterising causal relationships (Woodward, 2010, p. 304). The definition for specificity as influence is:

> **(INF)** There are a number of different possible states of C ($c_1 \ldots c_n$), a number of different possible states of E ($e_1 \ldots e_m$) and a mapping F from C to E such that for many states of C each such state has a unique image under F in E (that is, F is a function or close to it, so that the same state of C is not associated with different states of E, either on the same or different occasions), not too many different states of C are mapped onto the same state of E and most states of E are the image under F of some state of C. This mapping F should describe patterns of counterfactual dependency between states of C and states of E that support interventionist counterfactuals. Variations in the time and place of occurrence

[3] Lewis speaks of a 'joint cause' when he modifies the Billy-Suzy example to such a degree that both Billy's throw and Suzy's throw change the outcome significantly (Lewis, 2000/2004, p. 90).

> of the various states of E should similarly depend on variations in the time and place of occurrence of states of C. (Woodward, 2010, p. 305)

Subsequent to this definition, Woodward develops some thoughts on how to use it as a means of differentiating between the role of the DNA and the RNA polymerase during transcription. Woodward thinks that

> the functional relationship between DNA and RNA sequences is such that one can alter the latter in very specific ways by altering the former; but this is not so for the relationship between RNA polymerase and RNA sequence. The role of RNA polymerase in RNA and protein synthesis instead seems more *switch-like*. (Woodward, 2010, p. 306, italics in original)

While there appears to be a clear idea of how the DNA could be changed (by substituting nucleotides), I am not sure how Woodward intends to deal with the RNA polymerase. A view that appears to be consistent with Woodward's explanation is the following: the many possible combinations of nucleotides C $(c_1 \ldots c_n)$ exert a more fine-grained control over the messenger RNA E $(e_1 \ldots e_m)$ than the activity of the RNA polymerase which is either there or not there C (c_1, c_2). It appears that Griffiths et al. (2015) read Woodward that way, too, which indicates to me that it is a natural interpretation.

It is important at this point to understand why this is an idea that is much too vague and cannot be coherently spelled out. In a later passage of Woodward's article, he draws attention to the fact that it is possible to split variables or to lump them together (Woodward, 2010, p. 312). This highlights the problem that the choice of how one defines a variable will determine what one can measure and discover. Woodward also clearly states that it must be an empirical question of whether the RNA polymerase acts switch-like or not (Woodward, 2010, p. 307).

Consequently, variables have to be chosen very carefully if one doesn't want to miss something. One obvious thing to do is to define as a variable the concentration of RNA polymerase in a solution that contains DNA, RNA nucleotides, transcription factors and anything else needed for transcription. One can then change the concentration of the RNA polymerase and observe the transcription rate. Other things being equal, one would expect a near linear dependence between RNA polymerase concentration and transcription rate within certain limits. This appears to be what Woodward has in mind (Woodward, 2010, p. 306).

I am, however, not sure whether this is a suitable choice of variable, and I have two objections: the first is that other interventions are better suited if one wants to examine an effect on the RNA sequence, and the second is that even if one accepts Woodward's intervention it still gets associated with the wrong variable.

Although Woodward highlights the importance of treating the whole issue empirically, there is a hypothetical component in Woodward's account of which I am sceptical (Sect. 7.4), but which is of some significance in this situation. It is at least in principle conceivable to manipulate the RNA polymerase in more than one dimension: by changing the concentration of RNA polymerase, or by interfering with the proofreading capability of the RNA polymerase.

RNA polymerases possess a proofreading capability similar to but much less reliable than DNA polymerase proofreading mechanisms (Alberts et al., 2015, p. 304). It is commonly assumed that organisms can do without a well functioning RNA polymerase proofreading mechanism, since transcriptional mutations are not heritable (Alberts et al., 2015, p. 243). Yet it is also known that proofreading of the RNA polymerase differs between species and that there are mutations in the RNA polymerase that change proofreading reliability (Esyunina et al., 2016). Esyunina et al. suggest that the enhanced proofreading capability in some species is likely to be a response to extreme environmental stress such as mutagenic chemicals or radiation.

Imagine—hypothetically—that one somehow interferes with RNA proofreading by either increasing or decreasing the accuracy of that mechanism. An example of such an interference could be the substitution of amino acids in the RNA polymerase.

Now compare two interventions: one is increasing and decreasing the concentration of RNA polymerase, the other is changing amino acids in the RNA polymerase. Which intervention is suitable if one wants to know how the RNA polymerase affects the sequence of the mRNA transcript?

I think that the second intervention is the relevant intervention. The danger at this point is confusing two things: the modulation of the *amount* of a signal and the modulation of what is *in* a signal. (The former is in this case also a signal. The crucial point is that the coding differs.)

The activity of the RNA polymerase is correlated with the amount of mRNA in the cell and the amount of mRNA is responsible for the amount of protein that can be synthesised. Since the functioning of the cell depends on how many proteins of various kinds are present in certain proportions, it is vital for the cell to be able to increase or decrease the amount of the respective proteins. Hence, increasing or decreasing the activity of the RNA polymerase (at the appropriate transcription units) modulates a signal. Yet this modulated signal is *not* the mRNA sequence. What is modulated is the *amount* of instructions that are forwarded to the ribosomes. This is analogous to workers who carry out various repetitive tasks receiving orders for one table, then eight chairs, eight red wine glasses and then yet another table so that everyone can be seated at a table. The amount of orders that go out to the ribosomes has to be coordinated.

A different question is whether the orders that are sent to the ribosomes are properly written. In the analogy I am using here, this could mean that one may reduce the attentiveness of the workers who write the orders by offering them vodka instead of water. Assume that the workers who write the orders are not completely intoxicated so that they still send out the same amount of orders. Now the problem is that the workers who have to carry out these orders cannot decipher what is ordered: one unreadable thing of who knows what, eight thingamajigs, another eight of those things from which people drink . . . sherry glasses, maybe? . . . and finally one item of whatever that is.

These two modes of modulation must not be mixed up. Increasing and decreasing the concentration of RNA polymerase doesn't change the same signal as substituting amino acids in the RNA polymerase. If we now ask whether the RNA polymerase acts switch-like, we have either posed an imprecise question or begged the question. If it is imprecise, we should really have asked how the RNA polymerase changes the mRNA transcript. This would have required manipulating the amino acid sequence. In that case no switch-like relationship is found, and it must also be asked whether we are still speaking about specificity. It would seem that we are rather undermining the *stability* of the relation. Stability is in Woodward's framework the idea that causal relations continue to hold in changing background conditions (Woodward, 2010, pp. 291 ff.). One can of course debate whether a change in the amino acid sequence is a change in the background conditions. Yet thinking of the relation as being specific in the sense of influence is also strange. It is not that the RNA polymerase has *little* influence, but *none* at all. When the RNA polymerase matches the RNA nucleotides with the DNA template, the DNA is in total control of the message written. All the RNA polymerase can do is deteriorate the transmission. It is not that the RNA polymerase can in any meaningful sense exert control over the message in the mRNA. Hence, manipulating the amino acid sequence changes the mRNA transcript, but it is questionable whether we should still speak of specificity.

The other option is changing the concentration of RNA polymerase. In that case one doesn't see any effect on the mRNA transcript's sequence. This, however, is owed to the choice of variable and the associated intervention. It is *not* an empirical consequence. It was already predicated when that variable was chosen, instead of choosing a variable that affects the mRNA transcript's sequence. Deducing from this that the RNA polymerase acts switch-like on the mRNA transcript begs the question. Any tool acts 'switch-like' if you take it away from what it is supposed to do. That doesn't tell you which capability that tool has. Both hammer and screwdriver act 'switch-like' in this sense—which is a completely uninformative conclusion.

In short, if the RNA polymerase is manipulated in a way suitable for changing the mRNA sequence, it is likely that we are talking about stability instead of specificity, but if the RNA polymerase is manipulated by changing its concentration, we are no longer talking about the same effect.

What does this mean for Woodward's suggestion that his notion of specificity as influence may correspond to the idea of permissive and instructive causes in biology? I think there *is* an argument in favour of Woodward. This, however, requires going back to Holtzer's original argument, which was *not* about a comparison of RNA polymerase and DNA but about inducing molecules and nucleic acids. Holtzer argued that inducing molecules could not carry any significant amount of information, since induction can be triggered by a variety of simple, i.e. not very complex, chemical substances.

Such inducers are indeed engaged in a switch-like interaction. If they are present, they trigger a prearranged mechanism. If they are absent, the mechanism isn't triggered. One can also substitute such inducers with one another and always get the same result, i.e. the triggered mechanism if present and the dormant mechanism if absent.

The case for nucleic acids is different. If they are present, they trigger a mechanism, but they *also* provide input that changes the outcome. Nucleic acids can be substituted for one another, but this will not result in the same outcome. If we change DNA sequences or, further down the line, RNA sequences, this (provided it isn't a silent mutation) changes the protein produced at the ribosomes.

A comparison between inducers and nucleic acids in terms of Woodward's notion of specificity as influence is indeed meaningful: the inducers exert little influence, the nucleic acids exert great influence. This is in my opinion also very close to Holtzer's original idea that there is a difference between permitting and instructing substances.

Note that even Lewis's notion of influence applies! If one counterfactually changes inducers, this has little influence over the effect. If, however, nucleic acids are counterfactually changed, such changes exert a huge influence over the effect. This comparison is of course not made in a similar setting, i.e. an example of late preemption, but in a different setting where the processes compared to each other are not in any sense competing. Yet this doesn't matter. The competitive setting is no prerequisite for a meaningful comparison.

One has therefore to conclude that Holtzer's idea of distinguishing permissive from instructive causes is reasonably translated into the idea of influence, whether it be in Woodward's or Lewis's account.

A consequence is that this conceptual distinction, if it is easily translatable into interventionism or counterfactual theory, is unlikely to be an instance of pluralism. The terms 'permissive' and 'instructive cause' find a home in monist conceptual frameworks. Yet a crucial difference between them and the ultimate-proximate terminology is that they are not just empty shells but useful terms.

Next, let me complete this with an analysis of what went wrong in Woodward's attempt to advertise his notion of specificity. The problem is that you cannot compare the *input* to a mechanism with *parts* of the mechanism.[4] The RNA polymerase is a part of the mechanism called 'protein biosynthesis.' The most conspicuous input to this mechanism is the DNA template. In addition to this the mechanism gets input by other factors of which there are many especially in eukaryotic systems: The chromatin must not be methylised. Transcription regulators must not interfere by binding to the various DNA sequences, which, if bound by these factors, inhibit transcription. Other transcription regulators must, however, bind appropriately so that transcription can be initiated. The pre-mRNA undergoes the process of alternative splicing, i.e. introns are excised and decisions are made about which potential exons are also excised. Subsequent to this a down-regulation of the translation rate is still possible through RNA interference.

Note that the input falls into two categories: there is input controlling the type of protein produced and input controlling the amount of protein produced. While

[4] I am not going to provide an account of a mechanism and I am also undecided which notion of mechanism would be best suited, yet the notion of mechanism is firmly established in the literature, which should suffice for the moment. Accounts have been given by Glennan (1996), Machamer, Darden, and Craver (2000), Woodward (2002), and Bechtel and Abrahamsen (2005).

alternative splicing supersedes the input from the DNA template, other input has no control over what type of protein is produced.

This means that one must not make arbitrary comparisons. One can compare the input from the DNA template with the input that goes into alternative splicing, since both influence the same effect. One cannot, however, compare input down-regulating transcription with input changing the type of protein produced. One can instead compare input regulating the transcription rate with input regulating the translation rate, since both influence the same effect, i.e. the amount of protein produced.

Any comparison of the RNA polymerase in terms of specificity is meaningless, since the RNA polymerase doesn't provide any input to the mechanism. It is *part* of the mechanism, and processes input from different directions: it strings together nucleotides as instructed by the DNA template, and it engages in this task as often as instructed by transcription regulators. The RNA polymerase is not a source of input, *it processes input*. There are subtle details that do not allow one to treat the RNA polymerase as if it were a cause exerting influence.

5.3 Problems with Analogy Models

Calcott (2017) reads Woodward in the context of a debate on causal parity, i.e. on the question of how much two or more causes contribute to a single effect. This, in my opinion, is *not* central to Holtzer's permissive-instructive distinction. Holtzer compared *type* causes, i.e. his point was that inducers as general types exert less fine-grained control over their effect than do nucleic acids. There is no need to jump from there to comparing token causes against one another.

A key element in Calcott's argument is his discussion of an analogy Woodward draws between protein biosynthesis and a radio. The suggestion is that the role of the RNA polymerase corresponds to the on-off switch and that of the DNA template corresponds to the dial for choosing the station (Woodward, 2010, p. 307). I find this analogy unhelpful, as Woodward doesn't have a firm grasp of the protein biosynthesis which this radio is supposed to embody.

Having an ersatz discussion about the radio analogy is dangerous because it lacks nuance. A slight improvement to the radio analogy would be the addition of a volume control. The volume at which the radio plays could then be the analogy for the transcription rate. As an alternative, one could also rethink the variable for the on-off switch. The transcription rate is a function of time. The longer the RNA polymerase is active, the more mRNA is produced. Hence, the variable for the on-off switch should not take the values 'on' and 'off', but values of the *temporal duration of the on state*. In what follows, no such adjustments are made. The analogy is taken at face value, which means that there are problems ahead because the analogy is poorly chosen.

It would be much safer to have a discussion about the real thing, yet Calcott arranges his discussion around this example, probably because it is easily formalised as a mathematical model. The theoretical background is the common information theoretic approach by Shannon and Weaver (1949). Information in that sense is the freedom of choosing a message. That freedom is limited by the messages from which a choice is possible. The number of those choices can be written as the logarithm to the base of 2. In the case where there are only two choices this is $log_2 2 = 1$. If there are more choices, 16 for instance, this is $log_2 16 = 4$. In the first case the amount of information is 1 bit, derived from the expression 'binary digit', in the second case one has 4 bits of information (Shannon & Weaver, 1949, pp. 9 f.).

The flip side of the freedom of choosing a message is the uncertainty about which message is chosen. The uncertainty is greatest when all choices are equiprobable, and it is lowered when some choices become more probable than others. This uncertainty can be calculated as the entropy of the probability distribution of the choices (Shannon & Weaver, 1949, pp. 14 f.).

If one wants to connect this theory with Woodward's interventionist theory of causation, one has to translate 'choice' into 'the particular state of a variable'. This is what happens when Griffiths et al. (2015, p. 534) propose an entropy measure for the mutual information between cause and effect variables. Mutual information $I(E; \hat{C})$ is defined as

$$I(E; \hat{C}) = H(E) - H(E|\hat{C}) , \tag{5.1}$$

where $H(E)$ denotes how many choices or states there are in the effect variable and $H(E|\hat{C})$ tells one how many choices are left, given an intervention (hence the so-called do operator ' ^ ') on the cause variable. Mutual information is a measurement for the certainty that can be gained by knowing the value of the cause variable.

Griffiths et al. (2015, pp. 534 ff.) illustrate how this measurement applies in different scenarios. In a case where the relationship between the two variables is bijective, i.e. the effect variable takes a certain value if and only if the cause variable has taken a certain value, the information gained is maximal. No uncertainty is left and the remaining entropy is zero bits. Other setups are conceivable. Any state of the cause variable may result in any state of the effect variable.[5] In such a scenario the mutual information gained is lowest, i.e. knowing the state of the cause variable doesn't lower the remaining uncertainty. An intermediate case occurs when the value of the cause variable results in a range of values of the effect variable, where this range is smaller than the whole range of the effect variable. In such a situation, knowing the state of the effect variable lowers the uncertainty, but without approaching zero bits, since one cannot know the exact state of the effect variable. An inverse case (that leads to high certainty) is when certain values of the effect variable are set by a range of values in the cause variable. In such a scenario the amount of information gained is as high as in the bijective case. The difference is that more than one value of the cause variable allows one to conclude exactly in

[5] Griffiths et al. (2015, p. 535) speak of a 'limiting case', since this is no longer a causal relation as understood in an interventionist theory of causation.

which state the effect variable has to be. One should of course also expect that the former two situations can overlap, i.e. that a range in the value of the cause variable will result in a range of values of the effect variable.

Applying this account of causal specificity in the sense of gained mutual information, Calcott assesses the specificity of the on-off switch and the dial. Stipulating that Woodward's radio has an on-off switch and a dial for choosing between eight equiprobable stations, Calcott calculates that the specificity of the switch has to be 1 bit and the specificity of the dial 1.5 bits. Varying the number of stations results in different outcomes: 1 bit for four stations, 2 bits for sixteen stations or 2.5 bits for thirty-two stations (Calcott, 2017, p. 487).

Calcott calls this analysis 'competitive', since the two causes, i.e. switch and dial, are directly compared with each other in assessing the overall contribution they make to what the radio plays. This, however, isn't all that can be examined. Instead of comparing the contribution of dial and switch to a single effect, one may also—I would say has to—be interested in a second approach that measures specificities for other effects. Calcott calls this the 'hierarchical' approach (Calcott, 2017, p. 488).

In that approach the following effects are taken into consideration:

1. the effect of the dial on what we hear when the radio is on,
2. the effect of the dial on what we hear when the radio is off and
3. the effect of the switch on the dial.

In this hierarchical approach the switch is treated as a background cause controlling a foreground cause, i.e. the dial. Given that one is interested in the here-listed effects, the specificity of these causes is: 3 bits for the dial when the radio is turned on, 0 bits for the dial when the radio is turned off and 1 bit for the on-off switch (Calcott, 2017, pp. 488 f.).

Although this approach, too, assigns concrete values for the specificity of the respective causes, one has to emphasise an important difference, which is that the values of specificity are not comparable in the same way as they are in the competitive approach, where it is assumed that the two causes contribute to the same effect. One can still identify a cause that is specific to such and such a degree, but it is no longer a comparison between competitors for the same effect (Calcott, 2017, p. 489).

Calcott's concern at this stage of the argument is that the radio is too simple an analogy, but *not* for the reasons I outlined at the beginning! He argues that the radio doesn't invoke reliable intuitions about what the foreground and the background cause should be. This distinction is of course of no significance if one takes Holtzer's account as a reference point. It only appears to be relevant in some expositions of the subject matter, as, for instance, in Barresi and Gilbert's account.

But anyway, Calcott attempts to find a criterion for deciding whether the competitive or the hierarchical analysis is to be preferred and he does this by linking the issue to the idea of foreground and background causes. He argues that at first glance it may appear that the switch is clearly the background cause, but with more deliberation one may also judge that the dial could be a background cause when it is already set before the switch is flipped on (Calcott, 2017, pp. 489 f.).

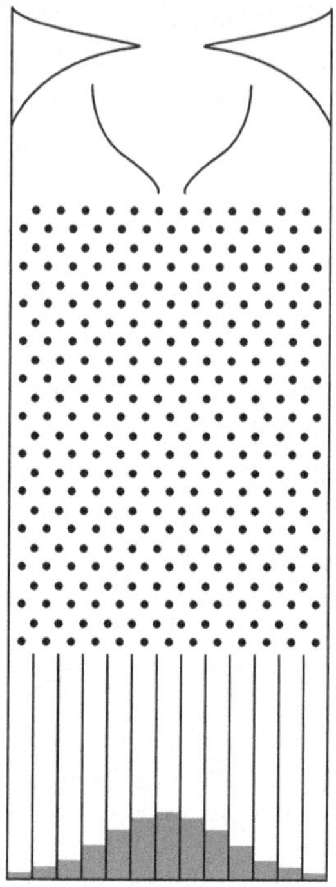

Fig. 5.1 Galton's box; redrawn from Galton (1894, p. 63)

Calcott's opinion is that it is impossible to decide which intuition to follow. In order to break the stalemate he switches to a new example, which means that he invents a new device called 'Waddington box'. That model is a merger of two other models: Galton's box for visualising normal distributions (Galton, 1894, p. 63) and Waddington's epigenetic landscape (e.g. Waddington, 1957/2014 or Waddington, 1977). In order to introduce Calcott's 'Waddington' box it will be sufficient just to speak about Galton's box and how the Waddington box is a variation thereof.

The essential part of Galton's box is a vertical board on which pins are arranged in many successive rows, one on top of the other (Fig. 5.1). The pins are placed in regular intervals and every second row is shifted sideways for half the distance of an interval. Below these rows are buckets that are also arranged in regular intervals with equal-sized openings. At the top, above the middle of the rows of pins, is a funnel through which marbles can be dropped into the device. On their way down, the marbles bounce off the pins. This diverts some marbles away from the middle. Hence the buckets at the bottom do not fill up equally. The buckets in the middle contain more marbles than the buckets at the periphery, thus creating a visual impression of a normal distribution (provided the front of this box is covered with a glass pane).

The Waddington box is, like Galton's box, a device through which marbles fall through a funnel, bounce back from pins and reach various buckets at the bottom. One important difference is that there is more than one starting point, i.e. there are two funnels or slots through which a marble can be thrown into the device. The other difference is that the pins can be arranged in changing patterns or layouts, i.e. the pins need not be uniformly distributed as in Galton's box (Calcott, 2017, pp. 490 ff.).

Depending on the chosen pattern of pins, marbles get directed towards different buckets, meaning that there are various probability distributions of how often the respective buckets catch a marble. One can try getting a marble into a bucket that lies further to the left or right by choosing the left or right slot as an entry point (Calcott, 2017, p. 493).

Calcott suggests that his Waddington box can be used as an illustration of how permissive and instructive causes differ. Choosing the left or right slot corresponds to flipping the switch of Woodward's radio. Rearranging the pin layout corresponds to turning the dial of Woodward's radio. What is different, according to Calcott, is that this device invokes strong intuitions about which cause acts instructively and which cause acts permissively. He also believes that he can demonstrate that 'permissive' and 'instructive' cannot mean the same as 'low specificity' and 'high specificity' (Calcott, 2017, pp. 492 ff.).

As before, Calcott conducts what he calls a hierarchical analysis, asking first what happens when one places the marble in either the left or the right entry and then how the layouts affect what happens when a marble is placed left or right.[6] Calcott shows that there are various extreme cases of what can happen: some layouts direct a marble towards a single bucket, which changes depending on the entry point of the marble; other layouts direct a marble to the left or right, depending on the entry point, but not towards a single bucket; and there are also layouts that direct a marble into a single bucket no matter through which slot the marble was dropped (Calcott, 2017, p. 493).

From here on Calcott's analysis takes a peculiar turn. He claims that one should *not* think of the choice of slot as the permissive cause:

> Recall that the origin of the permissive-instructive distinction was the induction of a cell or tissue type in development. A permissive cause thus plays the role of a reliable developmental switch: when it is on, one cell type or tissue develops; when it is off, a different cell type or tissue develops. Of the four examples [...] only LAYOUT p_4 produces a reliable switch, mapping each slot to one and only one bucket. This suggests that, rather than thinking of the slots as a permissive cause in general, we should think of them as permissive only in certain layouts. (Calcott, 2017, p. 494).

That of course is not what one finds when one goes back to Holtzer (1968) as the common reference point of the debate. Holtzer compared substances such as D_2O

[6] There are a huge number of possible layouts to consider. The number of layouts gets bigger the more positions for pins are taken into consideration. Calcott's particular model of a Waddington box has 66 positions at which one can either place or not place a pin. The total number of layouts in such a box is excessively high, i.e. 2^{66}. By imposing spacing rules on how pins can be arranged in the box, Calcott drives this number down to about 3.3 million layouts. Nevertheless, the only practical way of determining how marbles would fall. through various layouts is setting up a computer simulation, which is how Calcott arrives at a count of how often a specific bucket catches a marble (Calcott, 2017, p. 492).

with nucleic acids. D_2O, ions or other not too complex chemical substances are paradigmatic for permissive causes, nucleic acids are paradigmatic for instructive causes. Once it was known that inducers could not carry any detailed information, this raised quite naturally the question of how the mechanism of induction operates instead. This does *not* mean that the receptor and the signalling chain linking it to the genes on the chromosomes is the permissive cause. It is the mechanism researchers aimed to discover *after they identified inducers as permissive causes.*

Needless to say, with today's knowledge one can now also develop a model of how permissive mechanisms act in general. But that is a *different project.* Yet Calcott blurs the boundaries. In his opinion a permissive cause cannot be the same as a cause with low specificity. This could not be the case since the layouts vary with respect to the control they allow: some allow placing the marble in a specific bucket, others undermine control completely. Calcott claims that the low specificity cannot discriminate between cases where there is no causal control or merely 'a noisy relationship between cause and effect' (Calcott, 2017, p. 494).

In my opinion this is not surprising at all. It is always possible to enhance, undermine or completely destroy a mechanism. This is what happens, for example, when the amino acids in the RNA polymerase are manipulated. In such a scenario one cannot reasonably ask how high or low the specificity of the RNA polymerase is. By manipulating the RNA polymerase, the *stability* of the relation between the DNA template and mRNA can be undermined, but one cannot in any meaningful sense say that the RNA polymerase exerts influence over the mRNA sequence. In the case of the RNA polymerase only the notion of stability is applicable. Calcott's Waddington box raises a related problem. When the relation becomes more and more unstable the causal relation is eventually broken down completely. Applying the notion of specificity to a relation that is greatly unstable or at some point gone is also meaningless.

But I don't see how this is justification for rejecting Woodward's analysis, in the manner Calcott does. He deduces that Woodward is wrong in assuming that 'permissive causes are simply low specificity causes' (Calcott, 2017, p. 494). I cannot see how this would follow, unless one is totally confused about what one is investigating. Calcott suggests that one can utilise Waddington's notion of canalisation and 'call a permissive cause a *canalised switch*' (Calcott, 2017, p. 495, italics in original). I shall talk about canalisation later but in this context canalisation simply means that the pins are arranged in a layout that produces reliable results, i.e. if the marble enters through the left slot it ends up in different buckets than one that enters through the right slot.[7]

Why would I want to call the switch itself the cause? Calcott's reason for assuming that the switch itself must be the cause is that the relation holds only in certain contexts.

[7] In my opinion Calcott misinterprets what Waddington. said about induction as it occurs in the epigenetic landscape. I will analyse what Calcott got wrong in Sect. 5.6.

I agree that permissive causes are only permissive in relation to a certain mechanism. An inducer exerts its influence only if it comes into contact with a compatible mechanism. But is that any different from instructive causes? How does a nucleic acid exert its influence when the molecular machinery of the cell isn't around? Both permissive and instructive causes rely on intricate mechanisms. There is no difference in this respect.

When a competent cell reacts to an inducer, the cell can be described as a system possessing a disposition getting triggered by the inducer. There are different ways in which dispositions can become manifest. In some cases dispositions become manifest when an antidote is removed, e.g. releasing sand from the gondola of a helium balloon will allow that balloon to rise. In other cases dispositions become manifest under certain triggering conditions, e.g. the presence of a deer triggers a wolf's hunting instinct. In none of those cases would one confuse the antidote or the trigger with the disposition itself. Inducers and nucleic acids trigger dispositions, but they are still different from those dispositions.

What Calcott analyses are in fact the dispositions of systems, such as the Waddington box. Although such an analysis may be legitimate under a different name and purpose, it is not an analysis of permissive and instructive causes. Some scholars, such as Mumford (2009), think of dispositions as causes. Others, such as Hüttemann (2013), argue against that view, pointing out that dispositions are contributors but not causes (Sect. 2.6). I tend to believe that Hüttemann's reasoning applies here, too. However, the decisive argument need not be whether one sides with Mumford or Hüttemann on this issue. Even if one takes Mumford's view, the disposition of the inducer is different from the disposition of the cell.

Calcott's cardinal mistake is confusing the permissive input to a mechanism with the mechanism itself. His analysis doesn't refute Woodward's hypothesis that one can describe permissive and instructive causes in terms of varying degrees of influence or specificity.

5.4 Limitations of a Second Notion of Specificity

Bourrat (2019b) has argued with Calcott over the question of whether it is preferable to conduct a competitive or a hierarchical analysis. Calcott's opinion, with which I agree, is that one has to choose the hierarchical analysis. Bourrat, by contrast, thinks that one can save the competitive analysis if one turns to Woodward's second notion of specificity. I am deeply sceptical of this thesis. In order to explain why that is, let me have a look at Woodward's other notion of specificity:

> Put very roughly, this second idea is that a causal relationship is specific to the extent that a single (type of) cause produces only a single (type of) effect and to the extent that each single type of effect is produced only by a (type of) single cause. A non-specific causal relationship in this sense is one in which tokens of several different types of causes produce (are sufficient in the circumstances for) the same effect (e.g., both smoking and asbestos exposure cause lung cancer) or a single cause (smoking) produces a number of different effects (lung cancer, heart disease). (Woodward, 2010, p. 308)

It is worth quoting this description of a second sort of specificity at length once you realise that neither Hume's nor Mill's name occurs a single time in this essay. Although the resemblance is strikingly obvious, Woodward doesn't recognise that these are familiar observations. One–one specificity is similar to Hume's original account of causal regularity, where cause and effect are joined in an exclusive relation. What is then described as a 'non-specific causal relationship' matches the criticism and refinement of the Humean regularity theory by Mill. The fact that smoking and exposure to asbestos are both type causes of lung cancer is an instance of the Millian plurality of causes principle (Sect. 2.1).

Woodward notes that examples for the one–one conception of specificity can be easily found in the medical and biological literature. I agree that examples of this sort of specificity are ubiquitous in the life sciences. The most prominent case, mentioned by Woodward (2010, p. 309) is enzyme-substrate interaction. Such interaction can be specific to a degree at which an enzyme will interact with only one substrate. The phenomenon is commonly explained with the lock-and-key model or the more refined induced-fit model (Koshland, 1958).

However, such a seemingly exclusive relationship is dependent on the surroundings, i.e. the undisturbed operation of the organism. As soon as a disturbance occurs the relationship can break down: for instance, when an organism gets poisoned. A common phenomenon is that a poison interferes with a supposedly exclusive relationship by imitating the substrate.[8]

That the idea of one–one specificity needs to be hedged in a meaningful way is also on Woodward's mind. His proposal is to specify a range of alternative effects of a cause. Once such a range has been specified, causes may be compared against each other. Some effects may be brought about by only one cause, others may be caused by more than one or many alternative causes. Understood this way, one–one specificity is also a matter of degree (Woodward, 2010, p. 311).

One possible application I can think of is a scale of how specific an enzyme is under physiological conditions. Not all enzymes are specific to the same degree, and they could be ranked accordingly. Although this looks as if it would be a sensible application, there are serious questions remaining.

There is a significant difference between Woodward's notion of specificity as influence and one–one specificity. Specificity as influence is a well defined concept, since the variables occurring in its definition are also well defined. These variables take values of temperature, salinity, pH, pressure, distance, volume, light absorption and so on. For such variables it is clear what their range is. They report the grade of a property. This doesn't mean that they need to be continuous. A variable for salinity, for instance, is discrete since the ions in a solution can in principle be counted and salinity cannot be increased by only half an ion. Distance of course is continuous, since there is always a smaller fraction conceivable. Variables for

[8] An example for such a case of poisoning is the metabolisation of ethylene glycol by the alcohol dehydrogenase to glycoaldehyde, which when it is further metabolised results in metabolite acidosis and the failure of various physiological processes (Battistella, 2002).

one–one specificity are different. They are like Frankenstein's monster, arbitrarily strung together, and you have to be lucky to endow them with life. What does that mean for a variable pair such as *causes of cancer* and *cancer*? Cancer, I presume, is simple. That variable takes the values {cancer, no-cancer}. But who can confidently list the causes of cancer, *including* the yet unknown ones?

The idea of specificity is of course also that one has a measure for how many values in the cause variable change the effect variable. This means that the cause variable cannot simply take values of known causes of cancer, it must also include any known type effect that isn't a cause of cancer. One might also perhaps give different weight to the values of the variables. Not all substances are equally powerful or equally abundant in the environment.

The only circumstances in which one is inclined to count how many type causes result in a type effect or how many type effects result from a type cause are scenarios in which one considers some sort of undisturbed or isolated system. It is sensible to ask how many different types of reactions are catalysed by a particular enzyme if there is no external interference. This results in a meaningful comparison with other enzymes in the same or a similar system.

As soon as exogenous substances are taken into account the concept becomes less clear and eventually meaningless. Perhaps there is a way of turning this idea into an operational tool for thought. It is more likely, however, that one will either capitulate to the fact that one does not know how many alternatives there are, or one will have to restrict the notion of one–one specificity in a way that is always at danger of trivialising the concept: if one excludes interfering factors, one–one specificity is high; if one admits interfering factors, one–one specificity is low.

Bourrat, like Woodward, is apparently not fully aware of how old and also how trivial the idea of one–one specificity is. Although he acknowledges 'that one-to-one specificity is, at least to some extent, an arbitrary distinction', he is determined to hedge it in a meaningful way (Bourrat, 2019c, p. 10). His solution is to confine the notion of one–one specificity to cases where the particular experimental setup constrains the alternative candidate causes and effects (Bourrat, 2019c, pp. 10 ff.). I don't want to categorically dismiss this plan, but it is difficult to see how this would justify the sort of comparisons across categories that I ruled out in Sect. 5.2.

5.5 Confusion About Causal Backgrounds

One–one specificity is formalised by Bourrat with a measure he calls 'variation of causal information' or also 'causal one-to-one (un)specificity between C and E' (Bourrat, 2019b, p. 6). The 'un'-part is due to the fact that one–one specificity in this sense is highest when this measurement is zero.[9] In order to account for the

[9] Bourrat and Calcott have had a rather curious exchange about this issue. Calcott responded with an analysis that explored how the two notions of specificity are related, concluding that Bourrat's

distinction between permissive and instructive causes, Bourrat uses a normalised version of this measurement that takes values from 0 (most specific) to 1 (least specific). He suggests that permissive causes score high on the normalised variation-of-information scale, while scoring low on the mutual-causal-information scale (which I introduced in Sect. 5.3). The instructive cause, in contrast, would score low on the normalised variation-of-information scale, while scoring high on the mutual-causal-information scale (Bourrat, 2019b, p. 10).

This analysis is, as Calcott (2019) in my opinion correctly observes, completely unmotivated. There is no explanation of why this is supposed to capture the actual usage that biologists make of the terminology. Calcott complains—and I agree—that Bourrat doesn't provide any serious semantic analysis of how the authors he listed (Sect. 5.1) apply the terminology. The source on which Bourrat relies most is an earlier edition of *Developmental Biology*, i.e. the source which is completely detached from Holtzer's original account. What Bourrat quotes from Holtzer (1968) is also insufficient as an explanation of what permissive and instructive causes are supposed to be. Following the Barresi-Gilbert view, in which the distinction seems to come down to background and foreground causes, Bourrat claims that the

> distinction points towards a more abstract distinction in the philosophy of causation between 'background' or 'enabling' conditions, on the one hand, and 'triggering' conditions, on the other hand, with background conditions being classically regarded as less specific than triggering conditions. Following this analogy, a permissive cause corresponds to a background condition while an instructive cause corresponds to a triggering condition. (Bourrat, 2019b, p. 2).

If it was that simple, one should of course ask why an analysis is needed at all. Another more serious objection is, however, that the assertion about general philosophy is false. Backgrounds are chosen for various reasons.

Mackie thinks of his causal field as a background of factors that are within their normal range (Sect. 2.1). What the effect of a deviation from normal range will cause (whether it has specific or unspecific consequences) is not what distinguishes factors that are relegated to the causal field from factors that are part of the inus condition. It is consistent with Mackie's account that a change in a factor in the causal field can have very a *specific* effect, i.e. potentially exerting *fine-grained control*. The reason why a factor is part of the causal field is that this rarely happens, i.e. that the factor as it is normal.

Eells argues that context unanimity results in greater descriptive power, a position that is rejected by Dupré and Cartwright for epistemological reasons. When one looks at Dupré's argument for randomised experiments, the reasoning is that there are nolens volens factors that one cannot control but of which one might hope that they will on average cancel each other out (Sect. 2.3). Again the criterion is not

---

specificity would turn out to be a normalised version of mutual causal information (Calcott, 2019, p. 3 f.). Bourrat (2019a) agreed that this connection exists, but also insisted that he himself had pointed to that connection in Bourrat (2019b) and Bourrat (2019c). In other words, Calcott's analysis would reveal nothing new.

specificity. It is consistent with these views that some background factors would exert fine-grained control if one were to change them.

There is no consensus on what background conditions are: one can think of them as factors within normal range (Mackie), as factors that have to be constant (Eells), or as background noise that scientists have to take into account by designing their experiments prudently (Dupré and Cartwright). As far as I can see, it is not for the reason that they are unspecific that background conditions are chosen, although some background factors can of course turn out to be that way.

While the above are principled objections from a wider philosophical perspective, there is another weighty objection following directly from Barresi and Gilbert's account of permissive causes. The background in question isn't unspecific. It provides a significant amount of instructions. The case of the decellularised heart is an excellent example. What Barresi and Gilbert call the 'environmental conditions' (Barresi & Gilbert, 2020, p. 99) could obviously be referred to as 'background conditions' in more general philosophical terms. Yet these background conditions do all the heavy lifting.[10] Barresi and Gilbert talk about the origin of instructions, not about how specific those instructions are.

However, the main issue on which Bourrat and Calcott disagree—and on which I have to side with Calcott—is whether the competitive analysis is meaningful. Bourrat believes that his two-dimensional analysis allows him to make a direct comparison of how much influence the switch and the dial have on the sound produced by the radio. As I have said already, the radio is a poor analogy. Much more important is analysing the protein biosynthesis itself. There is in my opinion no coherent idea of what it means to lump together the effect on the transcription rate with the effect on the mRNA sequence. Those are coordinated but separate signalling pathways. A competitive analysis without a clear conception of what the combined effect of them is can only be a useless formalism.

I can therefore only support Calcott's position. The hierarchical analysis has to be preferred over the competitive analysis. However, it is important to note that this doesn't rule out every competitive comparison. One can of course compare the input given by the DNA template with the input given to the splicing mechanism. They contribute to the same effect, i.e. the mRNA sequence. Likewise, comparisons between different inputs regulating the amount of protein produced are coherent. What is *not* coherent are the comparisons across categories.

What about the other model? Is there something the Waddington box can teach? From my perspective, it is not essential to employ the Waddington box in an argument against a naive competitive analysis. The fact that different arbitrary inputs to the protein biosynthesis must not be compared with each other follows already from a careful analysis of the protein biosynthesis. Yet another question is whether the Waddington box teaches us anything about permissive and instructive causes.

---

[10] The cardiomyocyte progenitor cells might form some sort of tissue without the extracellular scaffolding of a decellularised heart, but not a functional heart (Ott et al., 2008).

Whether this is the case or not requires going back to the original model from which it is inspired.

5.6 Waddington's Epigenetic Landscape

The epigenetic landscape model is an analogy created by Conrad Hal Waddington (1905–1975)[11] who promoted a view somewhere between orthodox[12] Darwinism and Lamarckism. Waddington argued that soft inheritance, i.e. the inheritance of acquired traits (Lamarckism), and hard inheritance, i.e. the inheritance of innate traits (orthodox Darwinism), both play a role in the development of organisms. He suggested that some traits may first occur as a reaction to environmental factors and that these traits may later be 'superseded by an internal genetic factor' (Waddington, 1942, p. 563). As an example Waddington cites the thickening of parts of the ostrich's skin in various places. Here it appears to be plausible to suggest that callosities can be a reaction to non-genetic factors, such as 'the crouching position of the bird' (Waddington, 1942, p. 563), which have later been superseded by genetic factors, since these callosities occur already in the embryo. Whether this is true is not at issue here. The reason for citing this example is to illustrate the broader perspective from which Waddington argued. To Waddington it was obvious that any capacity to respond to external factors or stimuli 'must itself be under genetic control', which means that the capacity of the ostrich to produce callosities is genetically controlled and that natural selection selected 'for a genotype which gave an optimum response' (Waddington, 1942, p. 563).

Against this backdrop Waddington then analyses the concept of canalisation. He argues 'that developmental reactions, *as they occur in organisms submitted to natural selection, are in general canalized*' (Waddington, 1942, p. 563, italics in original). Waddington supports this claim with several observations. He points out that tissues are sharply defined. This indicates that cell lineages are pushed either in this or that direction. They are not meandering on an intermediate course. He also notes that minor disturbances in organ development are often non-detrimental, resulting in a 'standard end-product' (Waddington, 1942, p. 563). None of this sort of canalisation is supposed to be 'absolute', but Waddington argues that there is a strong tendency discernible (Waddington, 1942, p. 563). Another of Waddington's observations is that there is a difference concerning the constancy of wild type

[11] Full name and dates of birth and death are taken from Schmidt and Hackethal (1998).

[12] It is known that Charles Darwin wasn't an exemplary Darwinist—hence the qualification. He admitted the possibility of soft inheritance alongside hard inheritance: '[o]n the whole, I think we may conclude that habit, use, and disuse, have, in some cases, played a considerable part in the modification of the constitution, and of the structure of various organs; but that the effects of use and disuse have often been largely combined with, and sometimes overmastered by the natural selection of innate variations' (Darwin, 1859/2008, p. 109). In the light of this one could of course also say that Waddington's position, which is outlined below, is closer to the original Darwin.

individuals and laboratory mutants. While laboratory mutants differ wildly from one another, wild type individuals don't show such a wide range of variation. Here Waddington claims that some sort of 'buffering of the genotype' has to take place, which would shield it 'against minor variations not only in the environment in which the animals develop but also in its genetic make-up' (Waddington, 1942, p. 564).

The claim that variation between wild type individuals is limited is of course a delicate matter. It evokes essentialist ideas, i.e. the position that Mayr described as emphasising 'discontinuity, constancy and typical values' (Mayr, 1982, p. 38). The surprise, as Amundson (2005) notes, is that Mayr approved of Waddington's concept of canalisation! Here is what Mayr writes in *The Growth of Biological Thought*:

> The work of Schmalhausen, Waddington, and Lerner has shown that its architecture provides the genotype with such a stable integration that certain components of the phenotype may remain unchanged during phyletic divergence. The underlying *canalizations* and regulatory mechanisms seem occasionally to remain virtually untouched during evolution, and this accounts for the sometimes quite unexpected stability of seemingly trifling components of the phenotype. (Mayr, 1982, p. 213, italics added)

Although the general idea of canalisation was accepted by Mayr, one should perhaps also note that Mayr's treatment of Waddington is ambivalent: further along in the same work Mayr speaks of Waddington as if he were a heretic. He complains that Waddington had been 'thoroughly familiar with selectionism' but employed essentialist arguments in evolutionary analyses (Mayr, 1982, p. 518).[13]

The reason why Mayr accepts the idea of canalisation is of course given by the context in which Mayr mentions Waddington. It is a discussion of Darwin's recognition of the fact that individual traits that are not widely shared are of lesser value to biological taxonomy. A reconstruction of the tree of life requires the identification of homologies, which was an idea lacked by pre-Darwinian systematists who didn't intend to reconstruct genealogical relations. Waddington's theory of canalisation connects with this issue quite naturally, since it is a way of accounting for homologies.[14]

Waddington illustrates his theory of canalisation with a cybernetic model. This model, called 'epigenetic landscape', visualises the process of canalisation during development. Waddington thought of development as a progressive system as opposed to a terminating system. The key characteristic of a terminating system is that it maintains a steady-state. The steady-state may involve change, but only in the sense that the same happens over and over again. He suggested that systems for thermoregulation or the maintenance of oxygen levels are paradigm examples of terminating systems (Waddington, 1977, pp. 98 ff.). In contrast, a progressive system may, but need not, develop towards an end-state. What distinguishes these systems is that they change (in a non-repetitive way) while they are in operation. Ontogenetic development is a paradigm example (Waddington, 1977, pp. 103 ff.). Waddington supplements this with another terminological distinction: a terminating

[13] For a detailed analysis of Mayr's attitude towards Waddington see Amundson (2005, pp. 209 ff.).

[14] This makes it even stranger that Mayr didn't anticipate homologous genes (Sect. 4.8).

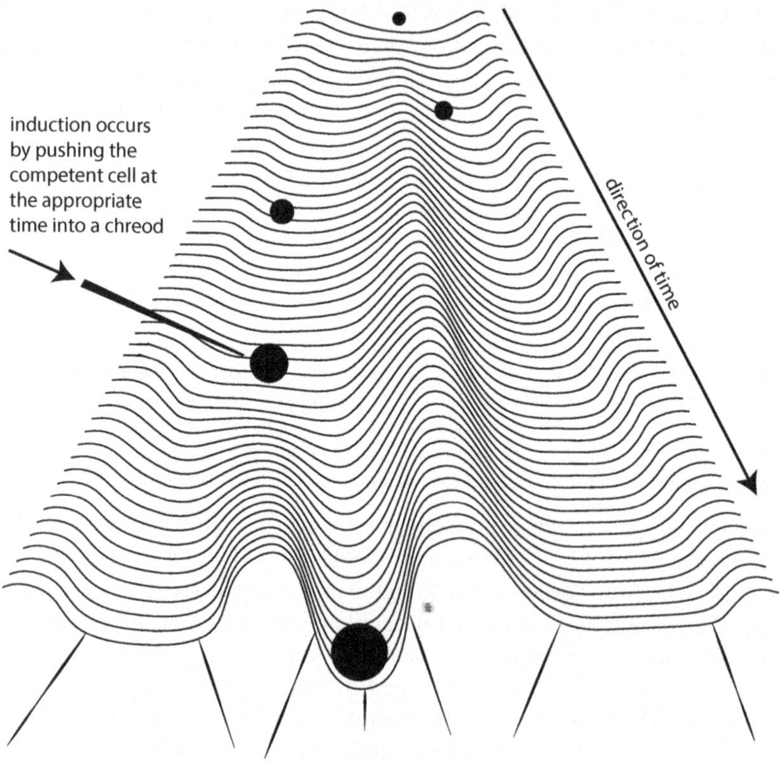

Fig. 5.2 My drawing of an epigenetic landscape in which I illustrate aspects of Waddington's idea. Similar (and more artistic) works can be found in Waddington's original publications

system maintains *homeostasis*, which means 'keeping something at a stable, or stationary, value', while a progressive system maintains *homeorhesis*, 'which means preserving a flow' (Waddington, 1977, p. 105).

The epigenetic landscape is a mental model that can be drawn as a picture (Fig. 5.2). The central element is an inclined surface such that a ball can roll from the vanishing point of that picture towards the spectator. The line of sight or the axis that points from the spectator towards the vanishing point represents time. The spectator is located in the present, while the vanishing point lies in the past. The ball represents a cell (or tissue) as it changes through time. Starting at the centre, the ball may roll downhill through valleys that split up as they get closer to the spectator. The different valleys that reach the bottom of the picture represent various end-states in which a cell (or tissue) may end up, or, rather, parts of the organism which are eventually formed by these cells (Waddington, 1977, pp. 106 ff.).

To the pathway set by a valley, Waddington gave the name 'chreod': a portmanteau word derived from two Greek roots, meaning 'necessary path' (Waddington,

1957/2014, p. 32).[15] Chreods have two significant properties: one is the change of the steepness of the path at the valley floor; which is called the 'chreodic profile'; the other is the profile of the cross section of the valley, which is called 'homeorhetic cross-section' or 'canalisation cross-section' (Waddington, 1957/2014, pp. 32 ff.). Waddington suggested that the epigenetic landscape might visualise different aspects of development and the phenomenon of canalisation in particular.

The chreodic profile roughly corresponds to the pace at which change takes place: in sections that are almost horizontal little happens, while in sections that descend steeply changes occur at a high rate (Waddington, 1957/2014, pp. 32 f.).

The homeorhetic cross-section can vary between two extremes: the chreod may have the shape of a gorge with steep walls, or it may resemble a flat and wide valley where a river might meander instead of following a straight path. Whether a chreod is of the former or the latter quality is reflected in the amount of variability that can be found in the traits of organisms: the flat and wide chreod will result in a high variability, the narrow gorge will force a trait towards a specific shared state of a trait.[16] How those chreods are shaped is determined by the genes pulling ropes from beneath the canvas which is the surface of the landscape (Waddington, 1957/2014, pp. 34 ff.).

The homeorhetic cross-section is part of the visual illustration of how disturbances during development are compensated. The ball, as it rolls through the epigenetic landscape, retains its momentum when it is pushed out of its normal path. This means that it won't return to the normal path at the point where it left but at a point that lies further down the road. Waddington suggests that this is what happens when an organism is injured during development: it will heal, but without going back to the stage during which it sustained the injury (Waddington, 1977, p. 106).

Waddington offers a similar account of how one may illustrate competence in this model. Competence is the phase during which a cell can respond to an inducer. In other words, the 'covertly differentiated cells' in Sect. 5.1 were competent cells. A competent cell in the analogy of the epigenetic landscape is a ball that can be pushed sideways such that it will roll down one specific valley instead of another. However, in order to succeed one must neither push too early nor too late when approaching the branching point between chreods. Waddington suggested that this corresponds to induction in real biological systems, i.e. that induction will only work during a specific time when cells are competent (Waddington, 1977, pp. 110 f.).

---

[15] There exist different spellings of this term. In Waddington (1957/2014) the term is spelt 'creode', in Waddington (1977) the spelling changes to 'chreod'. My criterion for choosing 'chreod' over 'creode' is that it occurs in the more recent work, presumably reflecting Waddington's change of opinion of how this term should be spelled.

[16] One of course has to assume that this variability isn't simply due to recombined alleles but also to the environment. The property of a phenotype that traits develop partly in response to the environment and not solely as a response to the genotype is known as 'phenotypic plasticity'. A related but more specific term is 'reaction norm' (Stearns & Hoekstra, 2005, p. 158 ff.).

Calcott's Waddington box changes certain aspects of the epigenetic landscape. One obvious change is that he raises the incline of the landscape so that the ball is now falling vertically through the model instead of rolling over its surface. This means that the ball can no longer be guided by the valleys' walls but has to be deflected by the pins in the box. The chreods, which are valleys in the epigenetic landscape, become shafts in the Waddington box, separated by the pins. One may of course question whether those shafts are still chreods. The odd thing about the Waddington box is that it has two entry points, while Waddington's epigenetic landscape has a single starting point.

In the epigenetic landscape the ball located at the vanishing point embodies the zygote. When it rolls downhill it embodies successive generations descended from the zygote, which get more and more specialised. The model of the Waddington box, in contrast, invites one to place the ball either left or right. What does that mean? The act of placing the ball is presumably the act of inducing cell fate. In the epigenetic landscape this is done by pushing the ball during a critical period, i.e. the time when the cell or tissue is competent. This temporal aspect of when to push is lost in the Waddington box.

Note that Waddington's metaphor accords well with the idea Holtzer had, i.e. that induction is caused by an external factor acting at the right time. Waddington does not conflate that external input with the surface of the epigenetic landscape. The surface of the landscape is determined by the genes, which means that the surface acts as a proxy for the instructional input from the genes. Calcott's analysis of a 'permissive switch' deals with this relation but not with the permissively acting cause, i.e the push of the ball.

Where Calcott (2017, p. 499) is right—I am now speaking about Waddington's epigenetic landscape rather than the Waddington box—is that manipulation occurs on different timescales. The push of the ball occurs during the lifetime of the organism. The change of the landscape occurs through mutation in germ-line cells during the lifetime of one organism affecting the life of a descended organism. What is *misleading* is that this should somehow bring out a difference between permissive and instructive causes. Whether a cause is permissive or instructive is a fairly intrinsic matter. The DNA template instructs here and now how to build a protein. The inducer permits here and now a change in gene expression.

Conflating this clear and unambiguous account with the temporal scale will only steer us back into the muddy waters of the debate on ultimate and proximate causes. It will illegitimately fuel the intuition that there are proximate causes permitting the release of genetic information and ultimate causes instructing how proximate causes act. We have been there, we have seen why that is wrong and we don't need to return to that place.

One should instead emphasise other aspects of Waddington's epigenetic landscape. It provides, although only by analogy, an explanation of how traits may evolve through an interplay of environmental stimuli pushing development within a chreod in a certain direction and a subsequent optimisation of the response through changes in allele frequencies. The environment acting permissively allows the development of a certain trait; the alleles that were able to react readily to the environmental stimuli (having provided instructions for the organism) get selected. This is a counterpoint to Mayr's simplistic idea of how ultimate and proximate causes interact.

Waddington notes that due to the lack of an absolute scale the chreodic profile isn't properly quantifiable, but rather a means for qualitative comparisons (Waddington, 1957/2014, pp. 32 f.). However, this doesn't mean that there have been no attempts at quantifying the model (Fagan, 2012). Jan Baedke (2013) has provided an extensive account of the history of the epigenetic landscape, how its influence faded over time and how it has now been revived by researchers. Baedke distinguishes four areas of application in which the model has been brought back: use as a tool for visualisation, a means of transdisciplinary communication, creative use and use as a modelling tool (Baedke, 2013, p. 759). The epigenetic landscape is apparently more popular than ever before, and I welcome that.

5.7 Conclusion

Woodward initiated the debate by asking whether biologists, when they speak of permissive and instructive causes, mean roughly the same as philosophers when they speak of low or high influence. That conjecture proves to be true if one takes Holtzer's account as the reference point. Maybe Woodward shouldn't have chosen the radio as an analogy, since it is provocatively simple compared with living organisms; this may then have motivated Calcott to come up with a model so complex that it can only be handled with computer assistance. Complexity, of course, is a distraction. One–one specificity is also a distraction. What matters is a careful analysis of what is known about actual biological mechanisms. I hope that I have succeeded in arguing that the RNA polymerase must not be the target of analysis if one is speaking of permissive and instructive causes, since the RNA polymerase neither permits nor instructs but *processes* input instead. If one now asks what permissive and instructive causes are, the answer comes down to: they are those causal factors which provide input to mechanisms and which may be ranked according to their influence. That ranking is of course only possible if those factors contribute to the same effect. A ranking for a gerrymandered effect is meaningless. The mechanism need not be complex. There are mechanisms of varying degree of complexity—the cell is a mechanism constituted by many component mechanisms, but so is a computer or even an ordinary light switch. This analysis is in my opinion the most compelling, since it is not just consistent with Holtzer's discussion of induction but also with Waddington's model of the epigenetic landscape.

References

Alberts, B., Johnson, A., Lewis, J., Morgan, D., Raff, M., Roberts, K., & Walter, P. (2015). *Molecular biology of the cell* (6th ed.). New York/London: W. W. Norton & Company.

Amundson, R. (2005). *The changing role of the embryo in evolutionary thought. Roots of evo-devo.* New York: Cambridge University Press.

Baedke, J. (2013). The epigenetic landscape in the course of time: Conrad Hal Waddington's methodological impact on the life sciences. *Studies in History and Philosophy of Science Part C: Studies in History and Philosophy of Biological and Biomedical Sciences*, *44*(4, Part B), 756–773. https://doi.org/10.1016/j.shpsc.2013.06.001

Barresi, M. J. F., & Gilbert, S. F. (2020). *Developmental biology* (12th international edn.). New York/Oxford: Sinauer Associates/Oxford University Press.

Battistella, M. (2002). Fomepizole as an antidote for ethylene glycol poisoning. *Annals of Pharmacotherapy*, *36*(6), 1085–1089. https://doi.org/10.1345/aph.1A397

Bechtel, W., & Abrahamsen, A. (2005). Explanation: a mechanist alternative. *Studies in History and Philosophy of Biological and Biomedical Sciences. Special Issue: Mechanisms in Biology*, *36*, 421–441.

Bourrat, P. (2019a). *Being precise about precision and one-to-one specificity.* Manuscript. PhilSci-Archive. Retrieved from http://philsci-archive.pitt.edu/16580/ (Deposited and last modified 31 Oct 2019 05:56)

Bourrat, P. (2019b). On Calcott's permissive and instructive cause distinction. *Biology and Philosophy*, *34*(1), 1. https://doi.org/10.1007/s10539-018-9654-y

Bourrat, P. (2019c). Variation of information as a measure of one-to-one causal specificity. *European Journal for Philosophy of Science*, *9*(11). https://doi.org/10.1007/s13194-018-0224-6

Calcott, B. (2017). Causal specificity and the instructive–permissive distinction. *Biology and Philosophy*, *32*(4), 481–505. https://doi.org/10.1007/s10539-017-9568-0

Calcott, B. (2019). Further clarification on permissive and instructive causes. *Biology and Philosophy*, *34*(50). https://doi.org/10.1007/s10539-019-9700-4

Crair, M. C. (1999). Neuronal activity during development: Permissive or instructive? *Current Opinion in Neurobiology*, *9*(1), 88–93. https://doi.org/10.1016/S0959-4388(99)80011-7

Darwin, C. (1859/2008). *On the origin of species by means of natural selection or the preservation of favoured races in the struggle for life* (Reprint of the 2nd ed.; G. Beer (Ed.)). Oxford: Oxford University Press. (Original work published 1859)

Esyunina, D., Turtola, M., Pupov, D., Bass, I., Klimašauskas, S., Belogurov, G., & Kulbachinskiy, A. (2016). Lineage-specific variations in the trigger loop modulate RNA proofreading by bacterial RNA polymerases. *Nucleic Acids Research*, *44*(3), 1298–1308. https://doi.org/10.1093/nar/gkv1521

Fagan, M. B. (2012). Waddington redux: Models and explanation in stem cell and systems biology. *Biology and Philosophy*, *27*(2), 179–213. https://doi.org/10.1007/s10539-011-9294-y

Galton, F. (1894). *Natural inheritance*. New York: Macmillan.

Glennan, S. (1996). Mechanisms and the nature of causation. *Erkenntnis*, *44*(1), 49–71. https://doi.org/10.1007/BF00172853

Grainger, R. M. (1992). Embryonic lens induction: Shedding light on vertebrate tissue determination. *Trends in Genetics*, *8*(10), 349–355. https://doi.org/10.1016/0168-9525(92)90280-H

Griffiths, P. E., Pocheville, A., Calcott, B., Stotz, K., Kim, H., & Knight, R. (2015). Measuring causal specificity. *Philosophy of Science*, *82*(4), 529–555. https://doi.org/10.1086/682914

Hall, N. (2004). Two concepts of causation. In J. Collins, N. Hall, & L. A. Paul (Eds.), *Causation and counterfactuals* (pp. 225–276). Cambridge: MIT Press.

Holtzer, H. (1968). Induction of chondrogenesis: A concept in quest of mechanisms. In R. Fleischmajer & R. E. Billingham (Eds.), *18th Hahnemann symposium. Epithelial-mesenchymal interactions* (pp. 152–164). Baltimore: The Williams & Wilkins Company.

Hunter, K. E., & Hatten, M. E. (1995). Radial glial cell transformation to astrocytes is bidirectional: Regulation by a diffusible factor in embryonic forebrain. *Proceedings of the National Academy of Sciences*, *92*(6), 2061–2065. https://doi.org/10.1073/pnas.92.6.2061

Hüttemann, A. (2013). A disposition-based process-theory of causation. In S. Mumford & M. Tugby (Eds.), *Metaphysics and science* (pp. 101–122). Oxford: Oxford University Press.

Kirschner, M. W., & Gerhart, J. C. (2005). *The plausibility of life. Resolving Darwin's dilemma*. New Haven: Yale University Press.

Koshland, D. E. (1958). Application of a theory of enzyme specificity to protein synthesis. *Proceedings of the National Academy of Sciences of the United States of America*, *44*(2), 98–104. https://doi.org/10.1073/pnas.44.2.98

Lemmon, V., Burden, S., Payne, H., Elmslie, G., & Hlavin, M. (1992). Neurite growth on different substrates: Permissive versus instructive influences and the role of adhesive strength. *The Journal of Neuroscience*, *12*(3), 818–826. https://doi.org/10.1523/JNEUROSCI.12-03-00818.1992

Lewis, D. (2000/2004). Causation as influence. In J. Collins, N. Hall, & L. A. Paul (Eds.), *Causation and counterfactuals* (pp. 75–106). Cambridge: MIT Press. (Reprinted from *Journal of Philosophy*, 2000, *97*(4), 182–197; minor revisions have been made for consistency)

Machamer, P., Darden, L., & Craver, C. F. (2000). Thinking about mechanisms. *Philosophy of Science*, *67*(1), 1–25. https://doi.org/10.1086/392759

Mayr, E. (1982). *The growth of biological thought. Diversity, evolution, and inheritance*. Cambridge: Belknap Press.

Mumford, S. (2009). Passing powers around. *The Monist*, *92*(1), 94–111.

Ott, H. C., Matthiesen, T. S., Goh, S.-K., Black, L. D., Kren, S. M., Netoff, T. I., & Taylor, D. A. (2008). Perfusion-decellularized matrix: Using nature's platform to engineer a bioartificial heart. *Nature Medicine*, *14*(2), 213–221. https://doi.org/10.1038/nm1684

Saha, M. S., Spann, C. L., & Grainger, R. M. (1989). Embryonic lens induction: More than meets the optic vesicle. *Cell Differentiation and Development*, *28*(3), 153–171. https://doi.org/10.1016/0922-3371(89)90001-4

Schmidt, I., & Hackethal, S. (1998). Kurzbiographien und Portraits. In I. Jahn (Ed.), *Geschichte der Biologie. Theorien, Methoden, Institutionen, Kurzbiographien* (3rd edn., pp. 755–1000). Heidelberg: Spektrum Akademischer Verlag.

Shannon, C. E., & Weaver, W. (1949). *The mathematical theory of communication*. Urbana, Chicago: University of Illinois Press.

Stearns, S. C., & Hoekstra, R. F. (2005). *Evolution. An introduction* (2nd edn.). Oxford: Oxford University Press.

Waddington, C. H. (1942). Canalization of development and the inheritance of acquired characters. *Nature*, *150*(3811), 563–565. https://doi.org/10.1038/150563a0

Waddington, C. H. (1977). *Tools for thought. How to understand and apply the latest scientific techniques of problem solving*. New York: Basic Books.

Waddington, C. H. (1957/2014). *The strategy of the genes. A discussion of some aspects of theoretical biology*. London: Routledge. (Original work published 1957)

Woodward, J. (2002). What is a mechanism? A counterfactual account. *Philosophy of Science*, *69*, S366–S377. https://doi.org/10.1086/341859

Woodward, J. (2010). Causation in biology: Stability, specificity, and the choice of levels of explanation. *Biology and Philosophy*, *25*(3), 287–318. https://doi.org/10.1007/s10539-010-9200-z

Chapter 6
Distinctions Between Production and Dependence

Abstract This chapter analyses three variants of the idea of differentiating between productive causation and causal relevance. I argue that there is no known (persuasive) application for either of these accounts of causal pluralism in biology. I conclude that there is no pressure for biologists to move towards these pluralist accounts.

Keywords Causal dependence · Causal pluralism · Causal production · Causal relevance · Drift · Frequency-dependent selection · Phylogenetic inference · Two concepts of cause

Many authors acknowledge some contrast between two 'sorts' of causation (Sect. 3.2), these being production (also known as 'productivity') on the one hand, and dependence (also known as 'dependency', 'quasi-causation' or 'causal relevance') on the other (Hall, 2004; Dowe, 2000). The idea is particularly popular with authors who discuss problems in the philosophy of biology (Sober, 1984b; Sober, 1984c; Glennan, 2009).

In this chapter I am going to discuss three authors: Hall (2004), Sober (1984c) and Glennan (2009). I will conduct an extensional analysis of these accounts, as I have done in previous chapters. What will be seen is that the general idea of distinguishing production, on the one hand, from some sort of dependence or relevance, on the other, can take many forms. Although superficially similar, they are quite different (one could even say disparate) accounts. Table 6.1 gives an overview of the results of the extensional analysis that follows.

Alongside that extensional analysis, I will discuss why one would want to make these distinctions. I shall turn to Hall's account first, since Hall develops his thoughts alongside some clear, even plausible, examples. However, while I admit that there are cases where one might want to differentiate between production and dependence, this is meant to be *only* a general acceptance of that terminology. The question of whether there is any need to invoke this distinction in biology remains open.

K. Ehrenstein, *Causal Pluralism in the Life Sciences*,
History, Philosophy and Theory of the Life Sciences 25,
https://doi.org/10.1007/978-3-030-87942-6_6

Table 6.1 Production-dependence distinctions: differences between authors with respect to their chosen relata (i.e. universes of discourse) and the extensional relationship between their concepts

	Sober (1984c)	Hall (2004)	Glennan (2009)
Relata	Events & properties	Events	Events/processes & facts
Extensions	Disjoint	Intersecting	Disjoint
Variety	Pluralism	Pluralism	Arity pluralism
Number of concepts	Two	Two	Three

When it comes to biology, the idea of the productivity-dependence distinction takes two different forms: one is Sober's account, which is meant to be motivated by problems that arise in evolutionary biology; the other is Glennan's account, which is inter alia motivated by his search for a possible solution to a dispute over frequency-dependent selection. I shall argue that neither Sober nor Glennan succeeds in motivating their respective distinctions from the examples to which they apply their terminology. In other words: none of these problems from biology hinges on a conceptual distinction between concepts of cause and effect.

6.1 Hall on the Causes of Forest Fires

As already mentioned in Sect. 3.2, Hall's account distinguishes between two concepts of cause that intersect. There are relations between events that are in the extension of both C^2_{22} (dependence concept) and C^2_{24} (production concept). This intersection is possible because Hall's universe of discourse, i.e. the set of causal relata, is homogeneous: its elements are only events (and omissions, which are—strictly speaking—nothing).

Consider some examples. Hall's universe of discourse is:

$\mathfrak{D}$: the set of token events

The ordinary concept of a cause can be formalised as the predicate:

C^2_2: ① is a token cause of ②

Hall's two concepts of cause are represented by:

C^2_{24}: ① produces ②
C^2_{22}: ① is causally relevant to ②

Then let me define these two constants for the following examples:

a: the event that the lightning occurs
b: the event that the forest fire occurs

Suppose that there is a dry forest and that a thunderstorm is brewing; lightning strikes (cause), and the forest is set on fire (effect). In other words, the event a that the lightning occurs is a token cause of the event b that the forest fire occurs:

$$C_2^2 ab \tag{6.1}$$

Note that event a is also causally relevant for event b. Had the lightning not struck (a), the forest would not have been set on fire (b). Hence it is also true that the lightning is a causally relevant token cause of the forest fire:

$$C_{22}^2 ab \tag{6.2}$$

Furthermore, event a is also a causally productive token cause of event b, since being struck by a lightning bolt is a paradigmatic case of productive causation:

$$C_{24}^2 ab \tag{6.3}$$

The relation $< a, b >$ is in the extensions of C_2^2, C_{22}^2, and C_{24}^2. There are three unanimous causal judgements.

Now consider a second example: Suppose that the lightning strikes, but that there is also a meteorite that hits the forest only fractions of a second afterwards. Suppose also that the meteorite alone (i.e. a hot glowing rock fallen from the sky) would have sufficed to set the forest on fire.

Prima facie one should agree with the two judgements that

the event that the lightning strikes is a token cause of the forest fire

and that

the event that the meteorite hits the forest is a token cause of the forest fire.

Define the constant

a_1: the event that the meteorite hits the forest

and formalise the above two judgements as

$$C_2^2 ab \tag{6.4}$$

and

$$C_2^2 a_1 b\,. \tag{6.5}$$

However, once one reconsiders the situation, one could also reason that neither lightning nor meteorite is a cause of the fire. Had the lightning not struck the forest, the forest fire would have occurred because of the meteorite; and had the meteorite not hit the forest, the forest fire would have occurred because of the lightning. Presupposing a naive counterfactual theory of causation, it would follow that neither the lightning nor the meteorite is a cause of the forest fire. Put into formal terms, one could also maintain that

$$\sim C_2^2 ab \tag{6.6}$$

and

$$\sim C_2^2 a_1 b \tag{6.7}$$

are true.

This contradiction is obviously unsatisfactory. In order to avoid it one could resort to a conceptual distinction.[1] By proposing that the concept of a token cause (C_2^2) is ambiguous and that 'causal relevance' (C_{22}^2) must be distinguished from 'causal production' (C_{24}^2), the contradiction could be avoided. One could propose that both the lightning and the meteorite are productive causes, but that they are not causally relevant causes. By choosing that option, the truth of all four following judgements could be maintained without contradiction:

$$C_{24}^2 ab \tag{6.8}$$

$$C_{24}^2 a_1 b \tag{6.9}$$

$$\sim C_{22}^2 ab \tag{6.10}$$

$$\sim C_{22}^2 a_1 b \tag{6.11}$$

Once one disambiguates C_2^2 into C_{22}^2 and C_{24}^2 these judgements are no longer in contradiction. It is possible to have causal production without having causal dependence.

However, it is also possible to have causal dependence without causal production. Consider a third example. The following case is discussed by Hall (2004, p. 229):

> There was heavy rain in April and electrical storms in the following two months; and in June the lightning took hold and started a forest fire. *If it hadn't been for the heavy rain in April, the forest would have caught fire in May.* (Bennett, 1987, p. 373, italics in original)

It would seem that this example would force one to conclude that the April rain caused the forest fire in June, since the fire in June depends on the rain in April. The rain in April ensures that there is enough combustible material in June.

Define the constants

a_2: the event that the April rain occurs

[1] Of course this isn't the only option. One could also propose that it is actually a case of preemption (Lewis, 1986, pp. 171 f.).

and

b_2: the event that the forest fire occurs in June

and state this formally as

$$C_2^2 a_2 b_2 \, . \tag{6.12}$$

To be read as 'the event that the April rain occurs is a token cause of the forest fire in June'.

Bennett believes that this would be counterintuitive and argues that 'no theory should persuade us that delaying a forest's burning for a month (or indeed for a minute) is causing a forest fire' (Bennett, 1987, p. 373).

But Hall disagrees. He believes that Bennett confuses type and token causation. It would be counterintuitive to maintain that rain in general (a type) causes forest fires, but that wouldn't mean that a specific rain (a token) couldn't be among the causes of a forest fire. Hall believes that one should disentangle type and token causation. Once one has done that, one may draw a different conclusion: according to Hall, one has dependence without production (Hall, 2004, p. 253). This means one can disambiguate the judgement into

$$\sim C_{24}^2 a_2 b_2 \tag{6.13}$$

and

$$C_{22}^2 a_2 b_2 \, . \tag{6.14}$$

To be read as 'the event that the April rain occurs (a_2) is *not* a productive cause of the event that the forest fire occurs in June (b_2)' and 'the event that the April rain occurs (a_2) is causally relevant to the event that the forest fire occurs in June (b_2)'.

To sum up, Hall's account distinguishes two notions of cause within egalitarian token-event causation, which allows him to disambiguate apparently contradictory causal judgements. Although effective in this respect, one might keep in mind that the price for this solution is to give up the idea that there is an overarching concept of cause and effect. That is why it is an instance of causal pluralism.

Hall's specific variety of the 'production-dependence' distinction is rarely invoked in the philosophy of biology. Yet it is instructive to consider it in order to see more clearly how other accounts invoke a distinction that is in some respects similar yet at the same time different.

6.2 Sober on Genealogy and Genetic Drift

The first thing to notice about Sober's (1984c) distinction between a *token* concept of cause that deals with events and a *type* concept of cause that deals with properties is that it runs 'horizontally' instead of 'vertically'. Hall's distinction separates token causal relations (that often coincide), whereas Sober's distinction separates token causation from type causation (which rules out that there is any intersection between these concepts). The interesting part, where similarities can be found, is that Sober's idea of token and type causation roughly corresponds to the idea that one can distinguish productivity accounts from dependency or causal relevance accounts of causation.

Sober doesn't really elaborate what token causation is, but he points to two other authors in order to give an idea of how one might understand token causation: one is Salmon (1978), the other is Anscombe (1971/1993).

Anscombe is well known for her view that causes need not determine their effects and that genealogical relations are paradigmatically causal relations: 'everyone will grant that physical parenthood is a causal relation' (Anscombe, 1971/1993, p. 92). The thematic connection with biology is plain to see.

That Sober refers to Salmon's account is also understandable: recall that Salmon's paradigm example of a Y-type interaction is a hen laying an egg (Sect. 2.5). Although Sober only considers asexual cases (as will shortly be seen), Salmon's account covers both modes of reproduction. Asexual reproduction is basically a copying mechanism. This mode of reproduction is quite common and can be found in organisms such as bacteria, algae and fungi. A bacterium, for instance, simply copies its genome and divides itself into two daughter cells. Multicellular organisms such as fungi may reproduce themselves by means of spores, i.e. specialised cells that separate themselves from the parent through mitosis. This mode of reproduction results in an offspring that is identical to the parent. Put differently, asexual reproduction is a succession of Y-interactions and each individual is genetically identical to its parent. Sexual reproduction is more complicated. It is a mode of reproduction where the genome is not simply copied. In the sexual reproduction cycle the genome gets divided into non-identical versions through meiosis. These non-identical versions of the genome are then fused together with other versions of the genome (usually) from other parents. This mode of reproduction results in an offspring that is different from both parents (Stearns & Hoekstra, 2005, pp. 177 f.). In philosophical terms, the difference between the two modes of reproduction is that asexual reproduction is a succession of Y-type interactions, while sexual reproduction is a succession of alternating Y-type and λ-type interactions. Invoking Salmon's account is a natural choice when one is discussing biology.

It is less obvious, however, how Sober integrates ideas from Anscombe and Salmon into his own account. What seems to be clear is that Sober's (1984c) account talks of token events instead of token processes. Hence, it would appear that an extensional analysis of Sober's terminology should postulate this:

$\mathfrak{D}^S$: the set of events and properties
C^2_{220}: event ① produces event ②
C^2_{222}: property ① is a causal factor at the population level for property ②

Since Sober also considers negative causal factors, one might complement the list with a further concept that mirrors C^2_{222}:

C^2_{224}: property ① is a negative causal factor at the population level for property ②

While Sober doesn't say much about token causation, he goes into more detail about the probabilistic approach. Phylogenetic inferences rely on what Sober calls 'property causality', known by others as the 'probabilistic theory of causality'. Any particular genealogy is a token phenomenon. Yet Sober argues that those token phenomena can only be understood in terms of property, i.e. type, causality.

When one asks whether, of three species A, B and C, the former two or the latter two share a more recent common ancestor D, one has to assess the likelihood of the competing hypotheses, where likelihood is to be understood as 'the probability it confers on the observations' (Sober, 1984c, p. 418). The competing hypotheses are $(AB)C$ and $A(BC)$, which might be further compared against the null hypothesis: that all three species are connected through a single branching point, as opposed to the two branching points required for any hypothesis that postulates a recent common ancestor for only two of the three species.

Sober (1984c, pp. 417 f.) stipulates that an unnamed and temporally more distant common ancestor possessed a character in the state 0. The state of the hypothetical more recent common ancestor D is unknown. The state of the character in the species A and B is 1, while it is 0 in species C.

It is easy to see that under such preconditions $(AB)C$ is the most likely hypothesis. It would require that the new character state 1 originated only once, provided there is a common ancestor D of A and B. The alternatives would be to postulate at least two independent origins of that character state under the $A(BC)$ or the null hypothesis (single branching point from which all three species are descended).

Sober argues similarly, but not in the same way as I have done here. He points out that '[i]t is worth noting that the common cause explanation here selected, which involves positing species D, does not say what character state D in fact has' (Sober, 1984c, p. 419). From this perspective, my assumption that species D might carry the trait 1 is unwarranted or at least unnecessary, since the principle of common cause requires no more than a coincidence of improbable consequences (Sect. 2.3).

Although this example of phylogenetic inference is intelligible, one might be more dubious about other examples that Sober employs to speak about how token and type causality differ. One example is about an instance of asexual reproduction, i.e. an instance of a Y-type interaction between a mother and a daughter: Assume that there is an individual that possesses a single set of chromosomes (haploid) and that reproduces asexually. Suppose also that there is a gene locus that has two possible states, 0 and 1. Now consider an individual, whose allele state is 0. When this individual (the mother) reproduces, the offspring (the daughter) will most likely possess the same allele, i.e. 0. But sometimes the copying mechanism of organisms fails and a mutation occurs. This is improbable but not impossible. In the case under consideration the offspring inherits the allele 1 despite the fact that the parent possesses the allele 0 (Sober, 1984c, p. 408).

Sober uses this example as an illustration of his thesis 'that at the token level, causes don't have to raise the probability of their effects' (Sober, 1984c, p. 407). Yet the odd thing about this example is that it is susceptible to all the counterarguments that have been given for similar cases, such as Rosen's example of the improbable birdie—an example also mentioned by Sober. We may recall Salmon's scepticism that it would always be possible, in any circumstances, to restore positive statistical relevance. However, he had to retreat into the realm of quantum theory to justify his scepticism. Rosen's example by itself does not yet force us to relinquish the thesis that causes must raise the probability of their effects.

Once one considers how biologists think about mutations, many options for restoring positive causal relevance are conceivable. Biologists distinguish types of mutation according to their *causes*. A common distinction is the one drawn between induced and spontaneous mutations. The characteristic of a spontaneous mutation is that it appears to have no (obvious) cause. These mutations in the nucleotide sequence 'are generally assumed to be accidental' (Klug et al., 2016, p. 404). Strictly speaking, this doesn't mean that they are or need to be uncaused. Nonetheless, these mutations are supposed to be 'a result of normal biological or chemical processes in the organism' and they are supposed to occur quite often during DNA replication (Klug et al., 2016, p. 404). Induced mutations differ with respect to their causes. They 'result from the influence of extraneous factors' such as radiation from the sun or radiation from (radioactive) minerals (Klug et al., 2016, p. 404).

If the mutation in question is an induced mutation, it is obvious that the interference of some environmental factor such as radiation might have increased the probability of the mutation. But even if the mutation were spontaneous, this would not rule out the possibility of there being some unknown factor *within* the cell that increases the probability of the mutation. A so-called spontaneous mutation is only supposed to be uncaused from outside the organism. Whether there is any intrinsic cause inside the cell doesn't matter.

Another example used by Sober raises the same question of whether any particular token concept of cause has to be adopted. This time the context is sexual reproduction, i.e. a scenario in which there are alternating Y-type and λ-type interactions:

> A characteristic is introduced as a mutation and strong selection increases its frequency. Just as the trait reaches a frequency of, say, 0.4, the environment changes, and the trait is rendered neutral. Subsequently, the trait, *by chance*, goes to fixation. (Sober, 1984b, p. 296, italics added)

Sober argues that if one were asked 'what caused the characteristic to reach 100 percent, it seems perfectly correct to answer: random genetic drift.' He suggests that '[s]ampling error can be causally efficacious, even though it sometimes will interpose itself into a historical process so as to reduce the probability of the outcome it in fact produces' (Sober, 1984b, p. 296).

Note that Sober is right in suggesting that this is how biologists think about drift. This view is reflected in standard textbook accounts, where drift is treated as a *cause*. This can be seen exemplified most prominently in section headings. In Stearns and

Hoekstra (2005, p. 66) a heading simply states that 'Kimura proposed that random drift causes most evolutionary change in DNA'. Another heading in Klug et al. (2016, p. 694) is 'Genetic Drift Causes Random Changes in Allele Frequency in Small Populations'.

The theory of genetic drift is built on the Hardy-Weinberg principle. It specifies conditions under which allele frequencies remain unchanged from generation to generation. One might therefore think of it as an inertia principle, or, as Millstein and Skipper put it, '[u]nderstanding evolution as a change in gene frequencies [...] is understanding the ways in which populations deviate from Hardy-Weinberg equilibrium' (Millstein & Skipper, 1998, p. 27). A population at Hardy-Weinberg equilibrium is an ideal population that is (in theory) infinitely large; its individuals mate at random, there are no mutations, there is no migration from or into the population, and there is no natural selection at work (Millstein & Skipper, 1998, p. 26).

The Hardy-Weinberg principle dictates that the trait's frequency should have remained at 0.4 in the absence of selection. A deviation from that frequency demands an explanation. Given that there is—as stipulated—no selection at work, that mating occurs (presumably) at random, that there are no further mutations, and no migration, it would appear that drift and finite population size are the explanation of the trait's going to fixation. In this sense sampling error is supposed to be causally efficacious. The ideal population is an inertial system and the sampling error is a disturbance of that system and, therefore, the cause of the deviation from Hardy-Weinberg equilibrium.

Both the example of phylogenetic inference and the example of genetic drift raise the question of whether one has to have an opinion about token causation at all. It is difficult to understand why Sober, although vague about the details of token causation, devotes an essay named 'Two Concepts of Cause' to the issue and why he claims in a related essay 'that a bifurcation of causal concepts is required' (Sober, 1984a, p. 232). That there are type and token concepts of cause and effect is not at issue. The question is whether there is any need to introduce a token concept of a cause that need not raise the probability of its effect.

If (hypothetically) all token causes did—contrary to Anscombe—somehow 'determine' their effects or did—contrary to Sober—increase the probability of their effects, phylogenetic inference could proceed the same way as before. A trait that originates as a consequence of some micro process that is accompanied by a probability increase can't change the phylogenetic argument, since this argument doesn't care about those micro details in the first place. Similarly, one might ask whether the concept of genetic drift presupposes anything about micro details. It is again possible to argue that biologists don't need to have an opinion about underlying micro processes. That is apparently also Sober's view when he says that '[t]he notion of chance used in evolutionary theory is independent of the question of whether determinism holds at the microlevel' (Sober, 1984a).

A potential problem is that someone might argue from a reductionist perspective and question the validity of such macro-level explanations altogether. Such an argument, which would have implications far beyond biology, is elsewhere refuted by Sober (2010), where he argues that macro-probabilities can be real even if they supervene on micro-probabilities and that even a deterministic supervenience base would not change that conclusion.

Note that my point is *not* to argue that Sober should discard his conceptual distinction. There might be a convincing argument elsewhere. My point is that I don't see how biology plays a role in such an argument. I can't see how any motivation flows from the biological examples themselves. Biology leaves much of the detail open for discussion and there is no obvious need to turn to a particular conceptual distinction.

6.3 Glennan on Frequency-Dependent Selection

A more recent issue that has been dealt with by postulating a conceptual distinction of concepts of cause and effect is frequency-dependent selection. Stuart Glennan proposed that a distinction between causal productivity and causal relevance would allow one to 'look for a middle ground' in the dispute over the interpretation of evolutionary theory (Glennan, 2009, p. 326).

The central problem is as follows: According to the traditional view, also called the dynamical view, '[n]atural selection is a causal process at the population level' (Glennan, 2009, p. 326). This position competes with two other interpretations of evolutionary theory: the individualistic view, which holds that '[n]atural selection is a causal process at the level of individuals'; and the statistical view, which holds that '[n]atural selection is a purely statistical phenomenon at the population level' (Glennan, 2009, p. 326).[2] According to these latter two interpretations, natural selection would turn out to be epiphenomenal. This is the key issue in the debate. Proponents of the traditional view discard the latter two interpretations for that reason (Glennan, 2009, p. 326).

Glennan's account of two concepts of cause is supposed to open some room for a compromise in this dispute. In a nutshell, that compromise is to acknowledge that natural selection isn't causally productive, though it is nevertheless causally relevant. As I understand it, the distinction between 'productivity' and 'causal relevance' is central to Glennan's proposal: this conceptual distinction is meant to solve the conflict between the adherents of the traditional position and those of the statistical position. My aim in this section is to try to answer the question of whether this solution is likely to succeed.

According to Glennan, any event has two sorts of cause: productive events in which an object is doing something, and causally relevant facts (Glennan, 2009). Accordingly, there are two sorts of causal judgement: One sort invokes a production concept of cause, e.g.

> The bowling ball knocked over the pin (Glennan, 2009, p. 327).

[2] Glennan adopts this basic outline of positions from Millstein (2006).

The other sort invokes a relevance concept of cause, e.g.

The fact that the key has a certain shape is causally relevant to whether it will open the door (Glennan, 2009, p. 329).

One may reformulate the two judgements in $\mathfrak{L}$. In order to do so I stipulate the following:

$\mathfrak{D}^G$: the set of events (in which an object is doing something) and facts
C^2_{322}: ① is causally relevant to ②
C^2_{324}: ① produces ②
a_{301}: the event that the bowling ball touches the pin
b_{301}: the event that the pin falls to the ground
c_{301}: the fact the key has a certain shape
d_{301}: the event that the key opens the door

One may then restate the above judgements as

$$C^2_{324}a_{301}b_{301} \tag{6.15}$$

and

$$C^2_{322}c_{301}d_{301}\ . \tag{6.16}$$

To be read as 'the event that the bowling ball touches the pin (a_{301}) produces the event that the pin falls to the ground (b_{301})' and 'the fact that the key has a certain shape (c_{301}) is causally relevant to the event that the key opens the door (d_{301})'.

These two concepts are also distinct: one has relations between events in its extension, the other has relations between facts and events in its extension. However, this doesn't mean that the concepts are unrelated. Glennan believes that 'relations of productivity and relevance are genuine and intertwined aspects of any true account of the causal structure of the world' (Glennan, 2009, p. 331).

This means that a comprehensive causal judgement would always cite productive events *and* facts as causes.[3] According to Glennan, the 'canonical form' of causal judgements is

c causes e in virtue of f (Glennan, 2009, p. 330).

A canonical judgement relates at least three elements: two events (c, e) and one or several facts (f) about these events (Glennan, 2009, p. 330). If one accepts this proposal, the above judgements are all elliptical. For instance, the judgement 'the event that the bowling ball touches the pin (a_{301}) produces the event that the pin falls to the ground (b_{301})' is incomplete in the sense that one should add 'in virtue of the fact that the bowling ball has the right momentum'. Likewise the judgement 'the

[3] This statement is more cautiously presented in Glennan (2010, p. 364), where it is conceded that this may not be a universal account.

fact that the key has a certain shape (c_{301}) is causally relevant to the event that the key opens the door (d_{301})' is incomplete: a more comprehensive judgement would be 'the event that the key is inserted into the lock and turned causes the event that the door opens in virtue of the fact that the key has the right shape'.

At this point one might consider whether there is a third concept hidden in Glennan's account, which would be that of a comprehensive cause:

C^3_{300}: ① causes ② in virtue of ③

If this were a concept, it would only be intelligible in contrast to the concepts of causal relevance and production, similarly to the way the concepts of net and component effect are related (Sect. 3.3). Consequently, one could think of Glennan's account as another instance of arity pluralism. Whether it is arity pluralism is of course as debatable as any of the reconstructions I offer, since it is always a matter of interpretation. Nevertheless, attempting these reconstructions helps to carve out difference more clearly. Glennan offers a further variation of the rough idea that one can distinguish production from dependence or causal relevance. This particular view of how one should differentiate between and integrate these two notions of cause is what shall now be tested by examining a central example that motivates that view.

Consider Glennan's (2009, pp. 331 ff.) treatment of the following example involving frequency-dependent selection. *Sigara distincta* is a species of water bug. The body colour of these water bugs varies. If one were confronted with a simple case of natural selection, one would perhaps expect that a *certain* shade of body colour is more advantageous than another: for instance, the colour form that offers better camouflage. Yet selection in water bugs is not that simple.

The water bugs are eaten by a certain fish species (it is not specified which). These fish, however, have a curious preference. They do not eat the water bugs that are poorly camouflaged, but those water bugs of the most abundant colour form. It has been shown that the success of survival of *Sigara distincta* depends on the rareness of a body colour. The rarer the colour, the fitter a so-coloured individual is. Since the preference of the fish depends on the frequency of a colour form, which colour is rare and which abundant may change over time. In other words, there is no colour form that is better adapted simpliciter. The hypothesis that is supposed to explain this phenomenon 'is that the fish predators form a stereotypic searching image associated with the dominant color' (Glennan, 2009, p. 331).

It would seem that in this example the increased frequency of a colour form is the *cause* of its decreased fitness and vice versa. Why would this be a case where two concepts of cause are needed? Glennan argues that there is a causal, although no productive relation. Production would hold between events. Relevance, by contrast, would be a relation between facts and events. Glennan denies that increase of frequency or decrease of fitness are individual events (Glennan, 2009, p. 332).

He links this claim to the account of a mechanism by Machamer et al. (2000). In Glennan's view, that account emphasises 'the productive role of objects' (Glennan, 2009, p. 332). In other words, only if objects are involved can there be productive relations. Glennan believes the problem would be that 'the population is not (in this case at least) an individual object' (Glennan, 2009, p. 331 f.). He argues that the population couldn't be an object, since it wouldn't be properly located in space and time. A criterion for being located in space and time would be that the entity engages 'in particular activities at particular times and places' (Glennan, 2009, p. 333).

Some populations (such as the water bug population) would not fulfil this criterion, since they would as a whole be spread out and not engaged in collective activities. Other populations engage in collective activities such as migration, but this is not the case in this example. Glennan argues that only the individual members of the population would engage in activities such as floating, swimming, evading predators, eating or mating (Glennan, 2009, p. 333).

But unlike other authors Glennan does not conclude that frequency-dependent selection is a non-causal phenomenon. According to Glennan, population-level properties are causally relevant (in the case of frequency-dependent selection) since the colour type frequency of a water bug makes the event that a specific bug is being eaten more or less likely:

> [L]et *e* be a predation event and *f* be the fact that the colour phenotype of the bug has a certain frequency. The fact that the frequency has a certain value will indeed make it more or less likely that *e* occurs. (Glennan, 2009, p. 335)

Glennan takes this as evidence that the trait's frequency is *causally relevant* to the occurrence of the event *e*. The central point in Glennan's argument is that one could have manipulated that population-level property in order to figure out whether it is causally relevant (Glennan, 2009, p. 336). In other words, we may change facts and these facts can be causally relevant. The proof would be to manipulate a fact—more specifically to carry out an intervention (Sect. 2.4)—and to observe effects of that manipulation.

This manipulation test was previously proposed by Shapiro and Sober (2007) and it is a key issue in the debate. Glennan accepts this test, but only insofar as it proves the causal relevance of population-level properties. He rejects, however, the idea that the manipulation test would also show that population-level properties are productive (Glennan, 2009, p. 337). This is Glennan's compromise position.

It is a position that is supposed to sit between the traditional and statistical interpretations of evolutionary theory. This middle position avoids the conclusion that population-level properties are causally inert, but maintains that population-level properties are not causally effective in a productive sense.

Nevertheless, how is invoking the conceptual distinction between 'causal relevance' and 'causal production' supposed to be convincing both to adherents of the traditional interpretation of evolutionary theory and to adherents of alternative interpretations? Why should those who advocate the statistical *or* the traditional interpretation of evolutionary theory adopt Glennan's terminology?

My answer will be that the distinction between 'causal production' and 'causal relevance' will most likely not settle the dispute between the adherents of the traditional interpretation and of the statistical interpretation.

Just consider this argument discussed by Glennan: The main criticism put forward by the advocates of a statistical/epiphenomenal interpretation of natural selection is that natural selection is a *pseudo-process* (see Sect. 2.5 again). Advocates of the statistical interpretation of evolutionary theory such as D. M. Walsh argue

> that the process of selection is more like the motion of a shadow [...]. It is simply the consequence of the differential rates of distinct causal processes occurring *within individuals*. [...] Natural selection and changes in gene frequencies are not related as cause and effect; they are joint effects of a common cause. Natural selection does not cause changes in gene frequencies. (Walsh, 2000, p. 141, italics in original)

This view suggests that natural selection is a pseudo-process that is only a consequence of genuine causal processes that occur at an individual level. Metabolism or meiosis, for instance, would be proper causal processes (Walsh, 2000, p. 140).

Glennan's reply to this argument does *not* rely on his conceptual distinction: according to Glennan, Walsh's error is to conceive of natural selection as a process that is caused by an individual-level process, while natural selection is actually *constituted* by individual-level processes. In other words, Glennan contests the supposition that natural selection is *caused* by individual-level processes (Glennan, 2009, p. 336). It should be obvious that this particular argument does not rely on a distinction between 'production' and 'causal relevance'. It would seem that Glennan's argument is not in this specific context dependent on the conceptual distinction between 'productivity' and 'causal relevance'. The discussion of this argument offers no clue as to why one should adopt Glennan's terminology. All that is shown is that one *can* adopt that terminology. There is, however, a difference between a tenable position and an advantageous position. Glennan's account may be as good as the traditional or the statistical account. But is it also advantageous in virtue of the conceptual distinction between 'productivity' and 'causal relevance'? That doesn't seem to be the case.

Consider also the following argument by Larry Shapiro and Elliott Sober. They, too, argue that Walsh's argument is incoherent (Shapiro & Sober, 2007, pp. 249 ff.). But why? The problem, according to Shapiro and Sober, is that Walsh and other adherents of the statistical interpretation of evolutionary theory are attempting to hold fixed the micro supervenience base of the variable on which they are intervening.

Shapiro and Sober compare two approaches to the question of whether there is a direct causal influence of the variable X on the variable Y. The first case asks whether there is a common cause of X and Y, i.e. whether a third variable screens off X from Y. In the other case, the micro supervenience base of the variable X is considered.

One can apply Shapiro and Sober's comparison to the water-bug example: Assume that X represents the fitness of a certain colour form of *Sigara distincta* at t_1 and that Y represents the fitness of another colour form of *Sigara distincta* at t_2. In order to figure out whether X has a direct causal influence on Y, one needs to fix common causes of X and Y before intervening on X. If the relation between X and Y is epiphenomenal, that intervention should not change the value of Y. If, by contrast, the relation between X and Y is not epiphenomenal, one should observe a change in the value of Y. That would be the basic approach.

Examining the relation between X and Y requires other causes to be fixed, particularly common causes of X and Y. In the first case this isn't a problem. One may hold fixed the common cause and examine what happens when one intervenes on X. In the second case this is more of a problem. It is impossible to intervene on X, while simultaneously holding fixed the micro supervenience base of X. One would be forced to imagine what would happen. This, however, would be (even in theory) an incoherent approach to the problem, according to Shapiro and Sober (2007, p. 240).

For Shapiro and Sober, the principal fallacy of the adherents of the statistical interpretation is that they suppose that X should have causal powers *additional* to the micro supervenience base. Shapiro and Sober deny that this is a meaningful account of epiphenomenalism (Shapiro & Sober, 2007, p. 251). They think that the whole approach of 'fixing' the micro supervenience base in order to search for additional causal powers is ill-conceived. They call this alleged argumentative strategy by adherents of the statistical interpretation 'the master argument for epiphenomenalism' (Shapiro & Sober, 2007, p. 241). Again, the conceptual distinction between 'causal production' and 'causal relevance' is irrelevant. The debate is over the question of whether a certain sort of manipulation is acceptable.

As so often in a philosophical dispute, the opponents of this argument are unimpressed. Matthen and Ariew (2009) argue that it is Shapiro and Sober who are insisting on carrying out a forbidden intervention. These proponents of the statistical interpretation of evolutionary theory believe that the intervention envisaged by Shapiro and Sober is 'mathematically impossible' (Matthen & Ariew, 2009, p. 211). Why do Matthen and Ariew hold such a view? They argue that there would be 'no room for a manipulable intermediary between advantageous heritable traits and the increase in frequency that results from their being advantageous' (Matthen & Ariew, 2009, p. 211). In their view '[s]uch an intermediary would not be distinct from the increase in frequency' (Matthen & Ariew, 2009, p. 211).

In order to get the gist of this argument one has to make a tacit background assumption explicit: Matthen and Ariew are presumably presupposing a common notion of fitness. This would be *relative* fitness (Orr, 2009, p. 532). In such an account 'fitness' means 'the contribution of a genotype to the next generation compared to the contributions of alternative genotypes for the same locus' (Campbell & Reece, 2005, p. 464). In other words, there is a mathematical relation between fitness values. If I manipulate the absolute number of offspring of one genotype (e.g. by killing or adding individuals of *Sigara distincta*), I will at the same time change the proportions of each genotype in the population. For instance, the fittest genotype 1 produces a progeny of 100 offspring. As a matter of definition, this fittest genotype has the fitness value 100%. Let us also suppose that genotype 2 produces 50 offspring. That would be a fitness value of 50% (relative to the fittest genotype). Suppose, furthermore, that genotype 3 produces 25 individuals, which would be a fitness value of 25%. If I then intervened on the fittest genotype by removing 10 offspring, the proportion of genotype 2 and 3 in relation to genotype 1 would change despite the fact that genotype 2 and 3 would produce exactly as many offspring as before (in total numbers). The values would be 55.6 and 27.8% respectively. The fitness of genotype 2 and 3 would increase without them doing anything 'better' than before.

Glennan raises a similar objection to his own example, but concludes that the ambiguity of the notion of fitness allows one to uphold the argument regarding *Sigara distincta* (Glennan, 2009, p. 337). Hence, the argument is not yet over at this point. However, my objective is not to settle this dispute. My point is once again that the conceptual distinction between 'causal relevance' and 'causal production' is irrelevant in this context. We are again talking about the legitimacy of an intervention. But that question cannot be solved by making a conceptual distinction.

The upshot, then, is that Glennan introduces a further position into the debate, but it is not obvious how Glennan's disambiguation of the notion of cause should attract proponents of other positions to his position. Naturally, this doesn't mean that Glennan's position is incoherent or implausible. My point is rather that one should be cautious about advertising it as a position that is superior *in virtue of* a conceptual distinction. My conclusion is, therefore, that we are justified in doubting the utility of the productivity-dependency distinction (as far as the arguments discussed here are concerned). If there is a relevant context in biology in which we need to invoke Glennan's distinction, we are yet to discover it.

6.4 Conclusion

When one asks whether biology is in dire need of a distinction that differentiates some sort of production from some sort of dependence or relevance, the answer is: No. Although I acknowledge that there might be some general purpose for such a distinction, I am sceptical of it having applications in biology. Invoking a conceptual distinction of causal concepts might even distract from the underlying issues which can persist even when one distinguishes production from dependence or causal relevance.

References

Anscombe, G. E. M. (1971/1993). Causality and determination. In E. Sosa & M. Tooley (Eds.), *Causation* (pp. 88–104). New York: Oxford University Press. (Reprinted from *Causality and determination. An inaugural lecture*, 1971, Cambridge University Press)

Bennett, J. (1987). Event causation: The counterfactual analysis. *Philosophical Perspectives, 1*, 367–386. https://doi.org/10.2307/2214150

Campbell, N. A., & Reece, J. B. (2005). *Biology* (7th ed.). San Francisco: Pearson Benjamin Cummings.

Dowe, P. (2000). *Physical causation*. Cambridge: Cambridge University Press.

Glennan, S. (2009). Productivity, relevance and natural selection. *Biology and Philosophy 24*, 325–339. https://doi.org/10.1007/s10539-008-9137-7

Glennan, S. (2010). Mechanisms, causes, and the layered model of the world. *Philosophy and Phenomenological Research*, *81*(2), 362–381. https://doi.org/10.1111/j.1933-1592.2010.00375.x

Hall, N. (2004). Two concepts of causation. In J. Collins, N. Hall, & L. A. Paul (Eds.), *Causation and counterfactuals* (pp. 225–276). Cambridge: MIT Press.

Klug, W. S., Cummings, M. R., Spencer, C. A., & Palladino, M. A. (2016). *Concepts of genetics* (11th ed.). Boston: Pearson. (Global Edition).

Lewis, D. (1986). *Philosophical papers* (Vol. 2). New York: Oxford University Press.

Machamer, P., Darden, L., & Craver, C. F. (2000). Thinking about mechanisms. *Philosophy of Science*, *67*(1), 1–25. https://doi.org/10.1086/392759

Matthen, M., & Ariew, A. (2009). Selection and causation. *Philosophy of Science*, *76*(2), 201–224. https://doi.org/10.1086/648102

Millstein, R. L. (2006). Natural selection as a population-level causal process. *The British Journal for the Philosophy of Science*, *57*(4), 627–653. https://doi.org/10.1093/bjps/axl025

Millstein, R. L. & Skipper, R. A. (1998). Population genetics. In D. L. Hull & M. Ruse (Eds.), *The Philosophy of Biology* (pp. 22–43). Oxford: Oxford University Press.

Orr, H. A. (2009). Fitness and its role in evolutionary genetics. *Nature Reviews Genetics*, *10*, 531–539. https://doi.org/10.1038/nrg2603

Salmon, W. C. (1978). Why ask, "why?": An inquiry concerning scientific explanation. *Proceedings and Addresses of the American Philosophical Association*, *51*(6), 683–705. https://doi.org/10.2307/3129654

Shapiro, L., & Sober, E. (2007). Epiphenomenalism: The dos and the don'ts. In P. Machamer & G. Wolters (Eds.), *Thinking about causes* (pp. 235–264). Pittsburgh, PA: University of Pittsburgh Press.

Sober, E. (1984a). Common cause explanation. *Philosophy of Science*, *51*(2), 212–241. https://doi.org/10.1086/289178

Sober, E. (1984b). *The nature of selection. Evolutionary theory in philosophical focus*. Chicago: The University of Chicago Press.

Sober, E. (1984c). Two concepts of cause. *PSA: Proceedings of the Biennial Meeting of the Philosophy of Science Association*, *1984*, 405–424. https://doi.org/10.1086/psaprocbienmeetp.1984.2.192518

Sober, E. (2010). Evolutionary theory and the reality of macro-probabilities. In E. Eells & J. H. Fetzer (Eds.), *The place of probability in science. In honor of Ellery Eells (1953–2006)* (Vol. 284, pp. 133–161). Dordrecht: Springer.

Stearns, S. C., & Hoekstra, R. F. (2005). *Evolution. An introduction* (2nd ed.). Oxford: Oxford University Press.

Walsh, D. (2000). Chasing shadows: Natural selection and adaptation. *Studies in History and Philosophy of Science Part C: Studies in History and Philosophy of Biological and Biomedical Sciences*, *31*(1), 135–153. https://doi.org/10.1016/S1369-8486(99)00041-2

Chapter 7
Omissions and Conceptual Distinctions of Causal Concepts

Abstract This chapter discusses which theoretical accounts are best suited to deal with omissions or disconnections as they occur in biological feedback mechanisms. I argue that the disposition-based process theory has the resources to account for biological feedback and that it can also exclude bogus omissions as causes or effects since it confines itself to actuality. Alternatives, such as interventionist theory or counterfactual theory, fail to exclude the absurd and far-fetched hypothetical interferences. I argue that there is also no satisfactory solution to this problem within those theories. Since one strategy employed by adherents of counterfactual theory is to deny that there is any need to solve the problem of absurd hypothetical causal explanations, I also develop an argument as to why this problem must not be ignored. I show that ignoring this problem disconnects counterfactual theory from explanatory practices in the life sciences.

Keywords Actual disturbance · Causal explanation · Causal pluralism · Causation by disconnection · Causation by omission · Counterfactual theory of causation · Disposition-based process theory of causation · Hypothetical disturbance · Interventionist theory of causation

A considerable amount of biological theorising relies on the idea of feedback mechanisms. Lactose induces the production of β-galactosidase, a surplus of tryptophan represses tryptophan production, insulin is part of a feedback mechanism and ecology is all about feedback regulation. While philosophers understand what feedback is, they struggle at the same time to conceptualise it. Their concern is that one has to speak about things that are *not* there. Lactose can be absent, but that doesn't mean that there is—ontologically speaking—an omission of lactose. This chapter is about philosophical strategies for dealing with omissions. I shall discuss which accounts are best suited to the challenge. My test cases are the *lac* and the *trp* operon, and one example from ordinary life: the gardener who isn't watering the flowers. My preference is for the normality approaches, as they provide most of the equipment needed for dispelling worries about omissions. When I discuss interventionism I will return to the issue of gerrymandered variables, which has

K. Ehrenstein, *Causal Pluralism in the Life Sciences*,
History, Philosophy and Theory of the Life Sciences 25,
https://doi.org/10.1007/978-3-030-87942-6_7

already posed a problem (Sect. 5.4). The frustrating conclusion is again that there is no compelling method for choosing well-defined variables, although this is a key issue in the context of this chapter. A similar objection applies to counterfactual theory, which provides in my opinion only partially satisfying answers. The biggest problem I see is that the accompanying theory of causal explanation is completely out of touch with biological theory. Towards the end of the chapter, I will discuss the pluralist approaches by Hall and Glennan. I will consider Glennan's approach as a possible contender for Hüttemann's disposition-based process theory, but I retain my preference for the latter.

7.1 Five Reservations About Omissions

Various authors have raised a range of metaphysical reservations about omissions.[1] One reason to view omissions with suspicion is the idea that cause and effect are supposed to be contiguous. One can trace this idea back to David Hume who held the view that 'cause and effect must be contiguous in space and time' (Treatise, 1.3.15.3/SB 173). Similar claims are repeated in the contemporary literature. Mohan Matthen and André Ariew state, for instance, that '[c]ausal influence is transmitted from one physical entity to another in a spatiotemporally continuous manner—that is, by every point in between the two entities being affected' (Matthen & Ariew, 2009, p. 213). And Marc Lange says that '[i]t would strike many of us as mysterious—magical, even—for two billiard balls to interact without touching' (Lange, 2002, p. 7).

Omissions do not fit into this picture. Wherever an omission occurs it seems to disrupt the contiguity of cause and effect in time and space. If one accepts contiguity as a universal characteristic of causation, one cannot accept that omissions are causes or effects. The only (apparent) option for upholding the contiguity thesis without abandoning the idea that omissions can be causes or effects would be to suppose that omissions are entities that occupy certain regions in time and space. Then it would seem that they could be causes and effects, or could connect causes and effects, just as any other entity that floats around in time and space can. Let me call this

1. the contiguity problem.

But this, too, would be a suggestion that could be viewed with suspicion. Can one treat omissions as if they were *really* there, where that which they are not would have been? That seems to be a rather strange hypothesis. The idea of an omission or an absence is that it is something that is missing. But to say that something is missing is to say that it is not there. Metaphysically speaking, it seems that 'negative events, negative states of affairs, and negative properties do not exist' (Moore, 2009,

[1] I thank Anne Nospickel for a conversation that helped me to compile the list in this section.

p. 444). And to treat something that is not there as something that is there seems to be a bad idea: '[a]bsences are spooky things, and we'd do best not to take them seriously' (Lewis, 2004b, p. 283). This leads into the so-called 'problem of the missing relatum' (Lewis, 2004b, p. 281). If causation is a relation between cause and effect, it requires that both cause and effect exist. If either the cause or the effect does not exist, causation cannot be a relation. Lewis calls it 'the missing relata objection' (Lewis, 2004a, p. 77). I call this

2. the missing-relatum/spooky-entities problem.

Furthermore, there is the opinion (already encountered in Sect. 2.5) 'that we do recognise, on reflection, that certain cases of prevention or omission are not genuine cases of causation' (Dowe, 2001, p. 218). There is, as Phil Dowe maintains, an 'intuition of difference': We have 'causation', on the one hand, and 'quasi causation', on the other hand (Dowe, 2000, pp. 124 ff.; Dowe, 2001, pp. 217 ff.). The former is a production concept that upholds the contiguity thesis. The latter is a dependence concept, which is supposed to cover informal talk that permits one to view omissions as causes or effects. Dowe acknowledges that there is also a contrary intuition that causation by omission is genuine causation. He calls that the 'genuinist intuition' (Dowe, 2001, p. 218). But this is in Dowe's view not the decisive intuition.

A similar suggestion is to distinguish 'production' from 'causal relevance' (Sect. 6.3). Stuart Glennan holds the view that 'omissions can be causally relevant but not productive' (Glennan, 2009, p. 334). This means that omissions can enter into causal explanations, but they are not seen as productive causes that can *produce* effects. According to Glennan, production requires locality but causal relevance does not (Glennan, 2009, p. 328). One may refer to this as

3. the problem with the intuition of difference.

A further worry is that once one omission is admitted to be a proper metaphysical cause, numerous other omissions need to be admitted. If I admit that A's not watering a flower caused the flower to die, I need to admit that B's failure, C's failure, and eventually anyone's failure to water the flower would be a cause of the flower's dying. Or, to take up one of Lewis's examples, I could also say that right now 'we are being kept alive by an absence of nerve gas in the air we are breathing' (Lewis, 2004a, p. 100). The worry is that if we accept one omission as a proper metaphysical cause, we have to accept all sorts of far-fetched omissions as causes.

There are two perspectives from which Lewis's objection may pose a problem. The first problem is that the sheer number of omissions that one would have to accept as causes would be implausibly high. The second problem is that some of the omissions would seem to be absurd regardless of the total number of omissions. The former would be a parsimony argument maintaining that one should not postulate more omissions than are required. The latter would be a *reductio* argument. A *reductio ad absurdum* does not rely on a high number of omissions. The argument can proceed by unveiling a single absurd consequence. One could argue, as Michael Moore does, that it would be absurd to suggest that the non-trampling of the grass in the park today was actually caused by 'some particular non-elephant' (Moore, 2009, p. 445). I call the first of these two problems

4. the relevance problem

and the other

5. the absurdity problem.

Note that Lewis can *only* rely on the reductio argument (absurdity problem) and not on the parsimony argument (relevance problem): Lewis distinguishes qualitative from quantitative parsimony. The former posits that one must not postulate too many 'fundamentally different *kinds* of entity' (Lewis, 2001, p. 87, italics in original). The latter demands that there must not be too many 'instances of the kinds' that one has already postulated (Lewis, 2001, p. 87). Lewis says that he recognises 'no presumption whatever in favor of quantitative parsimony' (Lewis, 2001, p. 87). Hence, only the reductio argument, which is a kind of qualitative parsimony argument, is open to him. If one takes a strict approach, one cannot allow Lewis to allude to the implausibly high number of omissions. Assume, for the sake of argument, that there is a single case that suggests accepting an omission as a metaphysical entity. Then I could defend other omissions in the same way as Lewis defends possible worlds: Lewis argues that his 'realism about possible worlds is merely quantitatively, not qualitatively, unparsimonious. You believe in our actual world already. I ask you to believe in more things of that kind, not in things of some new kind.' (Lewis, 2001, p. 87) I could use an analogous argument: You believe in some omissions already. I ask you to believe in more things of that kind, not in things of some new kind.

A common response to one or more of the problems listed here is firstly to deny that omissions exist and secondly to argue that this doesn't matter since they are not needed as truth makers for causal explanations. Such a position has been defended by Lewis who maintains that it does not matter 'when the relata go missing', since the counterfactual analysis 'can do without any causal relation at all' (Lewis, 2004b, p. 283). Helen Beebee has also argued that 'absences can figure in causal explanations even though they do not cause anything' (Beebee, 2004, p. 293). Note that I totally agree with Lewis and Beebee in this respect, but that I shall nevertheless argue that their notion of causal explanation is far too broad.

7.2 Omissions in Biology (and Beyond)

Omissions play an important role in a multitude of mechanisms. Jonathan Schaffer has presented a variety of examples in order to prove that point (Schaffer, 2004). One of his examples is muscle contraction: The basic contractile unit of a skeletal muscle is the sarcomere. The sarcomere is composed of thick and thin filaments. Thick filaments consist of myosin, while thin filaments consist of actin and tropomyosin. The myosin can interact with the actin such that the thin and the thick filaments slide past each other, thereby shortening the length of the sarcomere. However, this interaction is usually inhibited by tropomyosin, which blocks the myosin-binding site of the actin molecules. This inhibition ceases to operate in the presence of calcium ions that can be released from another cell compartment (the sarcoplasmatic reticulum). In skeletal muscles the release of calcium ions is (in turn) triggered by incoming nerve signals, i.e. action potentials (Campbell & Reece, 2005, Sect. 49.6). Hence, the shortening of the sarcomere is (inter alia) caused by an omission: the tropomyosin's failure to block the myosin-binding site. Schaffer has argued that this example shows that 'disconnections are both ubiquitous and paradigmatically causal' (Schaffer, 2000, p. 288). A philosophical illustration of the case is shown in Fig. 7.1.

Schaffer has also argued that absences serve as causes and effects as well as causal intermediates. Since Schaffer has already discussed this issue from a general perspective, I shall confine my discussion to just two paradigm examples from molecular biology which substantiate Schaffer's position. The criterion for selecting my examples is that they cover all four permutations of how positive and negative relata can causally interact. Given that any causal relatum is either a cause or an

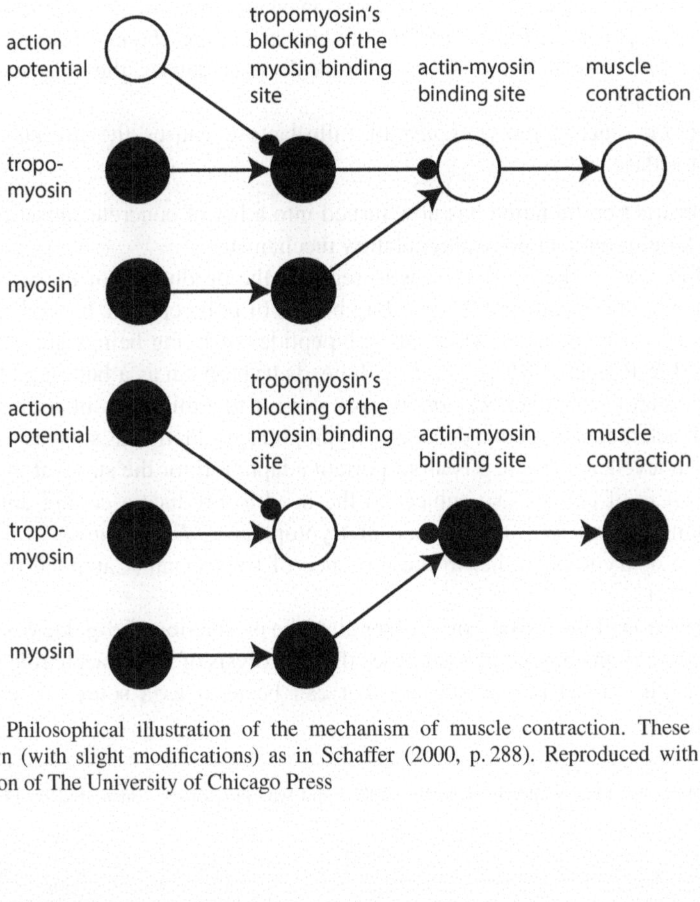

Fig. 7.1 Philosophical illustration of the mechanism of muscle contraction. These diagrams are drawn (with slight modifications) as in Schaffer (2000, p. 288). Reproduced with the kind permission of The University of Chicago Press

effect and either a presence or an absence, there are four cases to distinguish:

1. An *absence* causes a *presence*.
2. A *presence* causes an *absence*.
3. An *absence* causes an *absence*.
4. A *presence* causes a *presence*.[2]

The uncontroversial point on this list is item number four. All of the other three items raise the problems introduced in Sect. 7.1. The item that might be particularly suspicious is of course the third case. The odd thing is, however, that an example for this case was mentioned right at the beginning of this section: the absence of action potentials causes the absences of muscle contractions. Even more important is that this example is no strange exception. One can demonstrate this with two paradigm examples for genetic regulation: the *trp* operon and the *lac* operon in *E. coli*.

The *trp* operon is the typical textbook example for repressible operons. The *lac* operon is the typical textbook example for inducible operons. Both operons are representative of whole classes of mechanism that accord with the same basic principles (Campbell & Reece, 2005, pp. 352 ff.). These operons causally relate the presence and absence of certain substances in *E. coli*. The relations of interest are:

1. In the *trp* operon the *absence* of tryptophan causes the *presence* of tryptophan.
2. In the *trp* operon the *presence* of tryptophan causes the *absence* of tryptophan.
3. In the *lac* operon the *absence* of (allo)lactose causes the *absence* of β-galactosidase.
4. In the *lac* operon the *presence* of (allo)lactose causes the *presence* of β-galactosidase.

The abstract permutation has now turned into a list of concrete causal relations that are paradigmatic biological regulatory mechanisms.

The function of the *trp* operon is to regulate the production of the amino acid tryptophan. Amino acids are the building blocks of polypeptides. In total there are 20 kinds of amino acids of which all polypeptides of living beings are composed (Campbell & Reece, 2005, p. 79). *E. coli* needs tryptophan as a building block for its polypeptides. However, *E. coli* is also subject to limitations of its resources. The feedback mechanism that decreases tryptophan production as soon as sufficient tryptophan levels are reached is an important adaptation for the survival of *E. coli*. But *E. coli* is of course also subject to the need to produce a certain amount of tryptophan. This is why an absence of tryptophan in *E. coli* gives way to the production of tryptophan, i.e. to the presence of tryptophan (Campbell & Reece, 2005, p. 352).

The function of the *lac* operon is to regulate the production of β-galactosidase. β-galactosidase is an enzyme that catalyses the hydrolysis of the disaccharide lactose. Lactose (milk sugar) is one nutrient that can be used as a source of chemical

[2] This permutation also occurs in Schaffer (2000) and was previously considered by David Fair (1979).

energy by *E. coli*. However, lactose is just one of several nutrients and not always available. If there were no regulation of β-galactosidase production, there would be an excessive production of β-galactosidase in times of lactose shortage, which would be a waste of resources. It is for this reason that the presence of lactose *induces* the production of β-galactosidase, i.e. the presence of β-galactosidase. Yet, once the catabolism of lactose has succeeded, the absence of (allo)lactose[3] leads to the end of the production of β-galactosidase, i.e. to an absence of β-galactosidase (Campbell & Reece, 2005, pp. 354 f.).

Operons (no matter whether they are repressible or inducible) are sequences of directly adjacent genes that are united under the control of a single promoter (Campbell & Reece, 2005, p. 353). The promoter precedes a sequence of genes that code for certain polypeptides. The promoter itself has an internal structure. Two elements are crucial to its function: it has a recognition sequence that allows the RNA polymerase to locate the beginning of the operon, and it has a sequence that is called the 'operator'. The operator separates the recognition sequence from the genes that code for the polypeptides. The promoter's function is to allow a repressor to attach itself to the DNA. The repressor is a molecule that is constantly produced at a certain rate by a gene that lies outside the operon. Whenever a repressor is attached to the operator, the RNA polymerase finds the beginning of the operon but does not start transcription since the repressor blocks its way. Without transcription there is no mRNA copy of the DNA. And without mRNA the polypeptide that would have been produced during translation (the process that produces polypeptides by bonding amino acids) does not eventuate (Campbell & Reece, 2005, pp. 353 ff.).

While this description of the underlying mechanism applies to both repressible and inducible operons, there is a difference between repressible and inducible operons concerning the activation and inactivation of the repressor: if the operon is repressible, the repressor's default confirmation is inactive, which means that the repressor cannot attach itself to the operator unless it is activated by a ligand (Fig. 7.2). In the case of the *trp* operon the ligand is tryptophan. Whenever tryptophan attaches itself to the repressor, the repressor changes its confirmation into the active state. However, if the operon is inducible, the repressor's default confirmation is active, which means that the repressor can attach itself to the operator without activation by a ligand (Fig. 7.3). The binding of a ligand will inactivate the repressor instead. In the case of the *lac* operon this is done by allolactose (Campbell & Reece, 2005, pp. 354 f.).

If one looks at the details of how these operons work, even the fourth case (lactose causes the production of β-galactosidase) is no longer uncontroversial, since the disconnection of the repressor from the operator raises the contiguity problem.[4] It is also more than obvious that this raises the missing-relatum/spooky-entities problem. In my opinion, it is less clear whether this fuels the intuition of

[3] Allolactose is an isomer of lactose.

[4] Schaffer (2000) makes the same observation about muscle contraction and other examples, which all conform to the fundamental structure shown in Fig. 7.1.

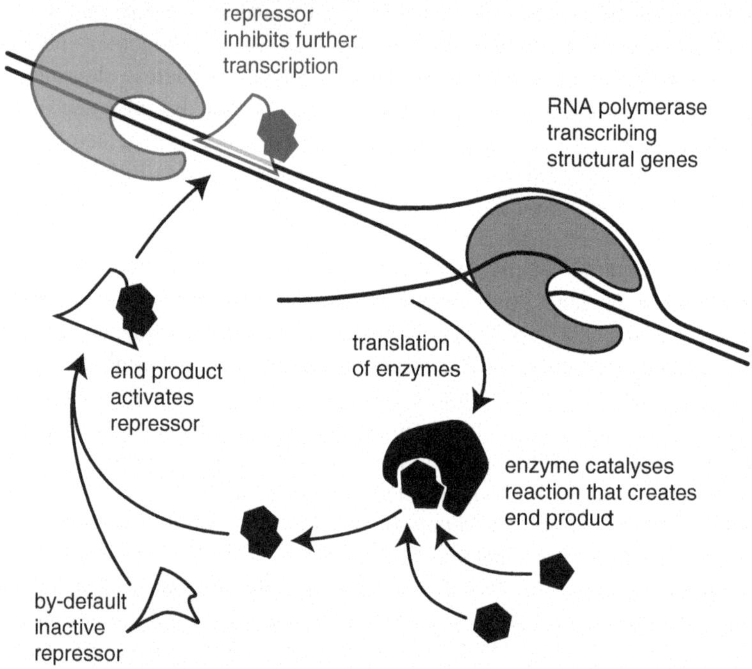

Fig. 7.2 My illustration of a mechanism of end-product repression

difference. I tend to side with Schaffer, so I would suggest that the *lac* operon is just another example which illustrates that '[c]ausation by disconnection is causation full force' (Schaffer, 2000, p. 289). However, the relevance problem is an issue. In all cases there are omissions that need to be distinguished from omissions that are not causally relevant:

1. Absences of tryptophan cause presences of tryptophan in the very same cell, but they do not cause presences of tryptophan elsewhere. Likewise, absences of tryptophan elsewhere do not cause presences of tryptophan in the cell, although they may cause presences of tryptophan elsewhere.
2. Presences of tryptophan cause absences of tryptophan in the very same cell, but they do not cause absences of tryptophan elsewhere. Likewise, presences of tryptophan elsewhere do not cause absences of tryptophan in the cell, although they may cause absences of tryptophan elsewhere.
3. Absences of (allo)lactose cause absences of β-galactosidase in the very same cell, but they do not cause absences of β-galactosidase elsewhere. Likewise, absences of (allo)lactose elsewhere do not cause absences of β-galactosidase in the cell, although they may cause absences of β-galactosidase elsewhere.

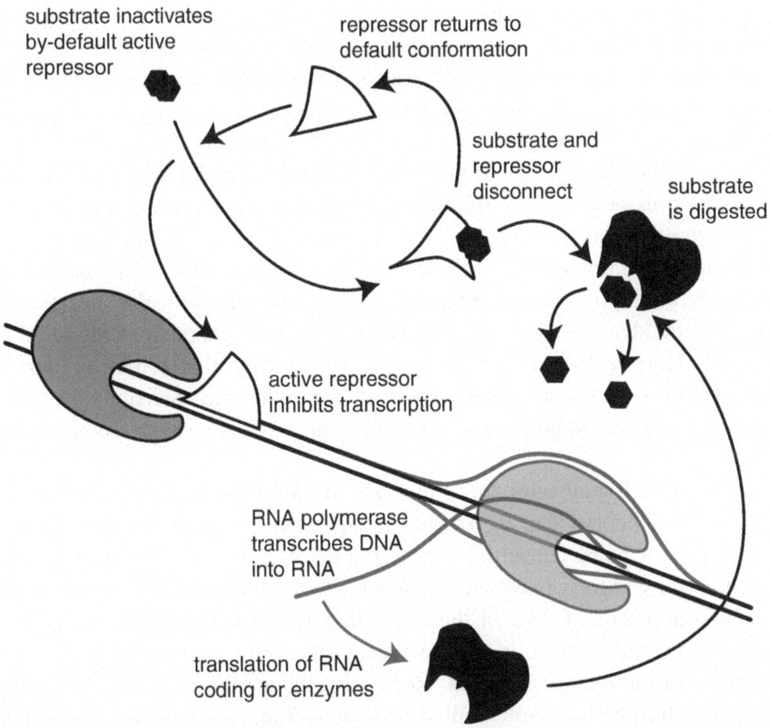

Fig. 7.3 My illustration of a mechanism of substrate induction

4. Presences of (allo)lactose cause presences of β-galactosidase in the very same cell, but they do not cause presences of β-galactosidase elsewhere. Likewise, presences of (allo)lactose elsewhere do not cause presences of β-galactosidase in the cell, although they may cause presences of β-galactosidase elsewhere.

Finally, there is the absurdity problem. Not only are there candidate relata that are far-fetched (relevance problem), but there are also those candidate relata that are plainly absurd (absurdity problem). It is far-fetched that a certain poison is not present. If the poison in question interfered with the above mechanisms, it would be sensible in some circumstances to point to the absence of such a poison as a relevant causal relatum (provided any such poison existed). It would be absurd, however, to point to the absence of things which we are quite certain do not exist or the existence of which we have no reason to suspect. One could, for instance, point to the absence of minuscular aliens (not composed of atoms and molecules as we know them) that hypothetically interfered with the above mechanisms when they travelled between the molecules of the *trp* operon or the *lac* operon. While it is a counterfactual truth that these aliens (if they existed) would interfere with the operons, it is absurd to suggest that their absence causes anything, given that one stipulates that there are no such aliens.

In order to connect with the more general debate, I will include one further example in the following discussion. Hart and Honoré (1985, pp. 38 ff.) considered how one would have to account for the death of flowers that occurs because of the gardener's neglect to water them. Variations of this example are often discussed in the literature (e.g. Beebee, 2004, p. 294; McGrath, 2005, pp. 125 f.). If the gardener fails to water the flowers, the flowers will shrivel up and die. But there are, of course, numerous other people (perhaps even strange aliens from outer space) that haven't watered those flowers. So, why does common sense suppose that it was the gardener's inaction that caused the flowers to die, rather than mine, or the chancellor's, or anyone else's? Discussing this example will also be instructive because it coincides—as Hart and Honoré note—'with a reprehensible dereliction from duty' (Hart & Honoré, 1985, p. 38). This means that critics of normality theories of causation (Sect. 2.6) will want to argue that any decisions taken in the evaluation of whose inaction it was must be subjective, or mere outcomes of moral prejudice.

Also note that I shall omit some theories of causation from the following discussion. Process theory (Sect. 2.5) virtually capitulates to the problem of omissions. Dowe's notion of quasi-causation is a desperate attempt at covering up the issue. Although I won't address Dowe below, most of what I have to say about Lewis and Beebee also applies to Dowe in the sense that Dowe would have to supplement his account of quasi-causation with a theory of causal explanation that respects explanations in the natural sciences. Regularity theory (Sect. 2.1) is also skipped, since it would have to be supplemented with ideas from Hart and Honoré, which are discussed in Sect. 7.3. Another theory family neglected is that of the probabilistic theories (Sect. 2.3). They are represented by their cousin interventionism (Sect. 2.4). Whether what I have to say about interventionism in Sect. 7.4 also applies to probabilistic approaches depends on the hypothetical element in them. Probabilistic theories that postulate causal relations between *measurable* variables are not vulnerable to the same criticism as interventionism.

7.3 Normality Approaches

If one ranks the theories in Chap. 2 according to their potential for dealing with omissions, it is obvious that the process theories in Sect. 2.5 are the least suited theories. All of the other theories portrayed in Chap. 2, with the exception of Hume's original position, have the advantage of not breaking down if contiguity is interrupted. The theories that might have an additional advantage are normality theories. One might expect that theories articulating—in Hüttemann's (2013) terminology—a dominant notion of cause would avoid relevance and absurdity problems better than theories articulating an egalitarian notion of cause. I will, therefore, begin with a discussion of how normality theories can cope with the challenges formulated in Sect. 7.1.

First I shall take a look at the problem of the shrivelled flowers. Hart and Honoré suggest that abnormal deviations are often omissions. In their opinion, the failure of a gardener to water the flowers may be seen as the cause of the flowers' death, since it had previously been normal for the gardener to water the flowers; but the failures of other people to water the flowers were not causes of the flowers' death, since it had not been normal for other people to water the flowers (Hart & Honoré, 1985, p. 38).

The problem is that Hart and Honoré's assessment of the situation relies on an intuitive understanding of normality (Sect. 2.6). It is, however, possible to support their argument by invoking concrete notions of normality. I would say that, on one hand, it is prototypically normal for gardeners to water flowers, and, on the other, it is also normal in a statistical sense. Invoking Schurz's (2001) connection principle, one can point out that gardeners who do not water flowers are selected against, in the sense that they lose their job. This explains why it is not just prototypically true that gardeners water flowers, but also why those gardeners are in the statistical majority.

What about other people? Can one argue that it is normal for a stranger or even the chancellor to water those flowers? If one assumes that the garden is private property, the first thing to notice is that there is selection against people who trespass on that property. This is a reason why strangers who water flowers in other's gardens are in the statistical minority. But if this is a public garden or park accessible to anyone, there is no selection against random visitors who don't water the flowers. Naturally, this is also not indicative of the contrary. In the absence of selection, drift acts unpredictably. Some flowers in public spaces are cared for by citizens,[5] other flowers in public spaces are neglected. If one is more specific about the person neglecting a flower, more can be said about why it is normal for them not to water a flower. Take the chancellor as an example: While there is selection against chancellors who crash the economy, there is no selection against chancellors who don't water flowers in the garden. There is, on the contrary, selection against chancellors who water flowers in gardens or parks, since this conflicts with the time-consuming duties for which chancellors are selected. What is true of the chancellor is probably also true of many ordinary citizens. They are occupied with other tasks such as going to work. That doesn't leave much time for watering flowers in the park. Hence it is also normal for the great majority of the population not to water flowers in the park.

Even some critics of Hart and Honoré, such as Helen Beebee (2004), admit that the example of the gardener and the flowers is one scenario which might be accounted for by normality theory. What Beebee nevertheless disputes is the possibility of generalising the normality approach. She refers to Jane Stapleton (1994) for a presumed counterexample: The scenario is that a dog contracts a disease which will eventually cause it to lose its eyesight. Although it is a potential cause of

[5] This is at least a phenomenon in Germany where some flowers in public space are planted and cared for by private citizens instead of public employees. These spaces are usually the open earth around trees lining the streets in the neighbourhood.

blindness, the disease is treatable, so the dog's fate can be averted if action is taken in good time. Unfortunately for the dog in this case, however, its owner ignores the disease.

Jane Stapleton argues that 'we would *probably* regard Z's conduct as a cause of the blindness because against the backdrop of Z's ownership of the dog we expect Z to have acted, not because of expectations generated by past conduct but for moral reasons' (Stapleton, 1994, p. 122, italics in original). In other words, because past conduct doesn't explain the causal judgement, she concludes that the dog owner's neglect is selected for moral reasons. The point of this argument is to show 'that it is not always true to say that "causal relation is a neutral issue blind to right and wrong"' (Stapleton, 1994, p. 122). Beebee sides with Stapleton and states that 'the abnormality criterion fails to explain our commonsense causal judgments' (Beebee, 2004, p. 295).

The problem with this argument is that it invokes a notion of normality that doesn't match Hart and Honoré's account. Although Hart and Honoré rely on an intuitive understanding of normality, it is not the notion of normality implicit in Stapleton's argument, i.e. the notion that what is normal is established by past conduct. Stapleton is arguing that it was normal for the dog's owner not to act because of how irresponsibly the owner acted in the past. If this were Hart and Honoré's position, it would be unintelligible for them to argue that the World Food authority's opinion about the cause of the famine is closer to the truth than the peasant's (Sect. 2.6). The World Food authority points to the government, the peasant to the prolonged dry weather. While the peasant might be judging the drought to be abnormal because of the sort of weather which occurred in the past, the World Food authority is certainly *not* judging the government by its past behaviour. If the government has failed to take precautions against the famine, that government is likely to have a bad track record already. This, however, would render the government's failure to counteract the famine *normal*. This cannot be how Hart and Honoré think of the World Food authority's reasoning. The World Food authority's reference point must be something else. The way I read Hart and Honoré, this must be the conduct of *other* governments. How other governments act is a matter of objective fact. Hence, the conclusion that the government's inaction is the abnormal factor that causes the famine is also objective and not a mere moral opinion.

This means that the common-sense judgement about the dog owner is also true. It is normal that dog owners care for their dogs. If you aren't satisfied with Hart and Honoré's intuitive notion of what is normal, try the statistical notion of normality. It appears that the statistical majority of dog owners care for their dogs.[6] The inaction of the dog owner considered by Stapleton is objectively abnormal.

A possible objection that remains is to complain about the inherent lack of intrinsicness in Hart and Honoré's account. Whether the gardener or the dog owner

---

[6] Since there are also many cases of animal cruelty worldwide, one could in principle overturn this judgement if one could show that the majority of dogs are mistreated by their owners.

causes harm to their charge is dependent upon the context. I agree that this can be a counterargument. One has to note, however, that it is a very general objection that also applies to any Humean theory of causality. Humean theoreticians should—by parity of reasoning—accept the defence given here of the common-sense judgement.

Those who are not tied to Humeanism may still ask whether an account of this common-sense judgement might also be found in the form of a theory that respects the idea that causal relations are intrinsic relations. Hence, next on the list is Hüttemann's (2013) disposition-based process theory. In order to apply this account, one needs to identify the subsystems that interact in a compound system and how the dispositions of these subsystems account for that interaction.

Each flower is a biological system that has the disposition to grow and flourish if it has soil, water and light. The gardener is also a biological system with certain dispositions. Since the gardener is an animal with highly advanced cognitive abilities, some of those dispositions were acquired through a learning process. The significant disposition in this scenario is the gardener's learned behaviour to water flowers. Each flower in the garden and the gardener himself are part of a compound system in which these individuals from different species interact.

Hart and Honoré say little about why the gardener didn't water the flowers. They only note that the gardener neglected his duty. However, a general knowledge of human beings tells us that there would have been a reason why the gardener didn't water the flowers. Possible disturbing factors or antidotes to the gardener's behaviour might be: occupation with other time-consuming tasks; preoccupation with thoughts that made him forget to water the flowers; or even an act of sabotage (in the event that the gardener holds a grudge against his employer). Whatever might have happened in particular, one can assume that it was an antidote to the gardener's disposition (learned behaviour) to water the flowers. The disturbance of the system called 'gardener' results in a further disturbance of the systems called 'flower'. With the soil and the light unchanged, it is the absence of water that acts as an antidote to the flowers' disposition to grow and flourish.

While this explains how the gardener and the flowers interact, it is not yet an explanation of how other people fit into the picture. One can also argue that for strangers it is a ubiquitous learned behaviour to water flowers, since many people (even if they are not gardeners) keep flowers on their window sill. Hence, one can argue that the neighbour has the disposition to water the flower but that there is an antidote to it, i.e. the fence around the garden with its closed gate. If one takes this view, there are of course as many causes of the flowers' death as there are people with the disposition to water flowers who cannot enter the garden. However, the flaw in this argument is viewing any stranger and the flowers as a compound system without any consideration. Hüttemann's account requires *actual* and not just hypothetical interactions of the subsystems. A stone lying on the ground, and the nearby lake in which the water would slow down the fall of the stone if it were thrown into the lake, are not a compound system. They become a compound system as soon as they interact, but not earlier.

Part of the challenge of applying Hüttemann's account is to identify the compound systems. How can one argue that the neighbour is not part of the compound system that keeps the flowers alive? It would seem that the neighbour and the flowers are not interacting, unlike the gardener who interacted with the flowers and then at some point stopped interacting. If the neighbour's disposition to water the flowers isn't triggered, there is no disturbance that acts as a cause of the flowers' death.

Whether or not the neighbour's disposition is triggered can depend on the circumstances. A possible scenario is that the neighbour can view the flowers across the fence. In that case one may argue that the neighbour's disposition to water the flowers is triggered in the sense that the neighbour feels an impulse to water these particular flowers. If this qualifies as interaction, the flowers and the neighbour become a compound system that can be interfered with. The interferences will usually be the fence or the fact that the flowers are on foreign property. One would then have to accept that it is not just the gardener's but also the neighbour's inaction that kills the flowers.

A further consequence is that passers-by who can see the flowers from the street and feel the impulse to water those flowers but refrain from it become part of the causal history of the flowers' death. This can increase the number of causes quite significantly if it is a busy street. What *cannot* happen is that the number of causes increases indefinitely. The number of passers-by are only a fraction of the world's population. This means that the account doesn't run into the relevance problem. That there is also no problem with absurd causes is obvious, since Hüttemann's account requires actual and no mere hypothetical interferences.

For the dog not receiving treatment a similar analysis can be given. Caring for a dog is a learned behaviour. Hence, the first question to ask is whether the dog owner learned how to care for his pet. That the dog owner didn't learn how to care for a dog is for most cases an unrealistic assumption. That is why people would usually see the dog owner's inaction as a cause of the dog's blindness. Even if the dog owner neglected the dog before, most people would not accept this as evidence that the dog owner didn't know how to care for a dog. They would—in my opinion correctly—assume that there must be an antidote to the dog owner's disposition to care for his dog. Depending on what the antidote is, the moral judgement of how much blame is laid upon the dog owner gets adjusted accordingly. If it is malice, such as a desire to see the animal suffer, the inaction is harshly condemned. If it is lack of money with which to pay the vet, the inaction is still seen as the cause, but it is morally excused and therefore not condemned so fervently. In cases in which it is plausible that the person didn't learn how to care for a dog, one indeed has to argue that the inaction of the dog owner wasn't the cause of the dog's blindness, since nothing could have triggered the disposition to care for the dog and take it to the vet. That, however, only applies to young children that have not yet developed that knowledge, or adults with a severe mental handicap. Note that most legal systems don't hold these people accountable for what happens around them.

Although one may extend this analysis further to other people who also didn't bring the dog to the vet, I shall leave the discussion here as it stands. It should be obvious that there can be a large (yet not a ridiculously high) number of people whose disposition was triggered but didn't become manifest. In all these cases one has to look at the antidote. Depending on what the antidote is, the moral judgement varies accordingly. The important conclusion I want to draw at this point is that critics of normality approaches are wrong on two counts: it is not true that normality accounts cannot single out dominant causes, and it is also not true that common sense regularly fails at that task. I see no reason to discredit *common* sense in order to push through any technical solution ignoring common-sense opinion.

At this point, I will now return to the two mechanisms which are paradigmatic for regulation in biology, i.e. the *trp* and *lac* operons. An interesting aspect is that the terms ‘repressible’ and ‘inducible operon’ can be read as conveying information about what is normal. These terms indicate the default configuration of the repressor molecule. The *trp* operon’s repressor is by default inactive, i.e. it is normal for that repressor not to bind to the operator. The *lac* operon’s repressor, in contrast, is by default active, i.e. it is normal for that repressor to bind to the operator.

The activation of the *trp* repressor occurs as soon as the level of tryptophan exceeds normal range. The abnormal level of tryptophan is then the cause of the return to a normal rate of tryptophan production.

The deactivation of the *lac* repressor occurs when lactose is present, which is not always the case and abnormal in the sense that *E. coli* doesn’t need to digest lactose. The digestion of lactose only happens when there is an opportunity. The (in this sense) abnormal presence of lactose causes the digestion of lactose. When the lactose has vanished, i.e. when the abnormal presence of lactose comes to an end, the repressor returns to its default conformation.

Although it is obvious that some notion of normality plays a role in these accounts of the *trp* and *lac* operon, one might of course still ask whether it is statistical, as I have suggested in the discussion of all the other examples. It wouldn’t harm *E. coli* to live in an environment with an abundance of lactose. Under such circumstances the high lactose level wouldn’t be abnormal, which means that the high lactose level could only be a condition. One has to ask, however, whether this is a plausible assumption about the circumstances under which *E. coli* evolved. The presence of this mechanism indicates that ancestors of *E. coli* probably endured prolonged periods during which they could not feed from lactose. Otherwise there wouldn’t have been any selection pressure to develop a regulatory mechanism that saves resources in the absence of lactose. One would then have to ask how this mechanism came about without selection pressure. For the *trp* operon a similar conjecture can be made. The normal range of the tryptophan level is very likely an outcome of natural selection. One can assume that cells that reduce tryptophan production at that specific level are in the statistical majority.

There is, however, still some strange consequence: since the absence of lactose is normal, an absence of lactose *cannot* be the cause of the absence of β-galactosidase. One would have to conclude that the absence of lactose is only a condition. In the tryptophan operon there is no such problem, since both the increase and the decrease of the tryptophan level are abnormal. Hence, a lack of tryptophan causes a return to the normal level but also an excess of tryptophan.

Note that the relegation of a candidate cause to the rank of a condition only occurs in the case where one absence results in another absence. Could this be indicative of an asymmetry between presences and absences? In order to explore this peculiarity I shall proceed by applying the disposition-based process theory once more.

What are the interacting systems? *E. coli* is the compound system in which various subsystems interact. The *lac* repressor, but also the enzyme β-galactosidase, are subsystems of that compound system. The *lac* repressor shows a default behaviour, i.e. being in a conformation that can bind to the operator, and binding to the operator when that default state is manifest. The presence of lactose is an antidote to the manifestation of that disposition. The repressor acted as an antidote to the manifestation of another subsystem, i.e. the RNA polymerase. With the repressor inactive, the RNA polymerase is now manifesting its behaviour of transcribing the structural genes in the operon. The mRNA created serves as a template for the production of β-galactosidase. The default behaviour of β-galactosidase is to catalyse the hydrolysis of lactose. Lactose, or rather its isomer allolactose, is disposed to bind to the *lac* repressor. The hydrolysis of lactose is therefore no disturbance preventing the default behaviour of lactose, but rather a disturbance annihilating the system that has shown the default behaviour of disturbing the *lac* repressor. When the lactose is digested, the level of β-galactosidase will eventually drop, since proteins degrade over time and are eventually recycled by the cell. Upon closer inspection, the absence of β-galactosidase is not due to the absence of lactose, but to other processes that degrade β-galactosidase. The production of β-galactosidase comes to an end, because the *lac* repressor can manifest its default behaviour after the antidote allolactose has been removed. It is the destruction of the antidote to the *lac* repressor's default behaviour which caused the end of β-galactosidase production, not the abstract absence of lactose.

According to this analysis, it is true that a lack of lactose subsequent to an excess of lactose leads to a situation where there is no longer any β-galactosidase around, although this is not true of the absence of lactose per se: the vanishing of lactose has an effect, whereas the absence of lactose (as the hypothetical assumption that there could be lactose where there is no lactose) does not.

This analysis is a bit more subtle than an approach in the spirit of Hart and Honoré. Since the disposition-based process theory demands that causes have to be actual disturbances, it points to the vanishing of lactose and not to the mere absence of it (as would be required in the Hart and Honoré account). What then is the asymmetry between presences and absences? It is—and here I agree with Beebee (2004)—a matter of existence and non-existence. However, not everything that is loosely called an 'omission' is an omission in the ontological sense. The 'omissions' that common sense easily accepts as causes are in fact *actual disturbances*. The destruction of lactose is just as concrete an event as the demolition of a house. One can say when and where these events take place. The not being there of lactose or the not being there of a house are indeed non-entities. That is why it is a virtue of the disposition-based process theory to exclude them as causes. The same applies to the effects that can be loosely referred to as 'absences' or 'omissions'. That there is no β-galactosidase is not what is caused in the strict sense. It is the end of β-galactosidase production that is caused. And that is a spatiotemporally localised event.

In this sense one has to subscribe to Schaffer's thesis about disconnections in causal theory. Yes, this is causation full force. It is, however, Hüttemann's theory that puts the relevant 'omissions' on an equal footing with positive events. If causes are disturbances of the default behaviour of *actual* systems, the distinction between positive and negative events becomes irrelevant.

7.4 Interventionism

I have argued that most of the problems surrounding omissions in causal theory can be avoided by a normality approach. Can the same be said of approaches that do not distinguish causes from conditions but aim to articulate a broad egalitarian or, in Lewis's terminology (Sect. 2.2), a non-discriminatory notion of cause? In other words, is it possible to analyse the notion of *a* cause as opposed to non-causes in a way that doesn't include irrelevant or even absurd omissions? I shall first discuss Woodward's interventionist theory and later Lewis's counterfactual theory.

In interventionist theory, events and omissions can be represented as values of indicator variables (Sect. 2.4). It is easy to see that such an approach is prone to include dubious variables, such as {*chancellor watering flowers*, *chancellor not watering flowers*}. However, it is also true that Woodward places restrictions on the choice of variables. He insists that variables have to be 'well-defined' (Woodward, 2003, p. 112). According to Woodward, a variable is *not* well-defined if it refers to a property 'that, for logical, conceptual, or metaphysical reasons, must be possessed by every object' (Woodward, 2003, p. 112). The idea of a 'well-defined' variable is further restricted by the requirement that the change needs to accord with a 'coherent idea' (Woodward, 2003, p. 113). What this means is illustrated with a counterexample: according to Woodward, it is incoherent to choose a variable such as '"animal" which takes the values {*lizard*, *kitten*, *raven*}' (Woodward, 2003, p. 113). Note that Woodward thinks of this variable as a change in *property*. He argues that there is no coherent idea which could govern the changing of a lizard into a kitten, or the kitten into a raven. It would not be enough to have a cage with changing content, i.e. with the lizard being taken out and the kitten put in. This, according to Woodward, would be a *different* variable that takes values of the content of the cage. Unfortunately, this is almost the entire explanation he provides of what it means to be well-defined. Woodward states that he relies on 'an intuitive understanding of what this notion involves' (Woodward, 2003, p. 113).

With such a rudimentary account it is of course difficult to proceed. I am also not sure whether I grasp Woodward's intuition. Although I agree that the variable that takes values of the content of the cage and the variable that takes values of the property of being an animal are different, I am not sure whether this helps to explain how Woodward wishes to exclude certain properties from being causes, such as 'being a member of a certain species, being a member of a particular race, and being a certain age' (Woodward, 2003, p. 113). The part of the account that is intelligible is the idea of tying the notion of being well-defined to the notion of an experiment. Hence, properties that can be changed in an experiment are properties that can be chosen as variables. However, Woodward doesn't confine himself to actuality: hypothetical experiments are also allowed (Woodward, 2003, p. 113, f.).

This hypothetical component makes it in my opinion difficult to understand why the variable 'animal' should be ill-defined. What is the difference between changing the animal in the cage as a whole and changing the animal in the cage (hypothetically) molecule by molecule? The practical complications aside, one can coherently imagine the transformation of the kitten molecule by molecule into the most recent common ancestor of kitten and lizard. That ancestor is then gradually transformed into the lizard. With the door to the realm of hypothetical experiments wide open, this is a perfectly coherent thought. Note that it is not even unusual for a variable to be changed by running through its values in a particular order. If one raises the temperature of an object from 283.15 K to 293.15 K, the temperature will rise by taking all the in-between values in a set order. As can be seen, the ramifications for rejecting a variable are rather vague. Are they strong enough to exclude the dubious cases?

The variables in the gardener example are of course G taking the values {*gardener-attends-to-flowers*, *gardener-does-not-attend-to-flowers*} and F taking the values {*shrivelled-up*, *not-shrivelled-up*}. If one intervenes on G, this changes the value of F. Naturally, one can introduce such indicator variables for many other people: neighbours, postmen, policemen, chancellors, and so forth. All these variables, were one to intervene on them, would change the value of F. Interventionism doesn't set any boundaries in these cases.

As a defence, one could argue that interventionism is a type-level theory. If one is not considering concrete, i.e. spatiotemporally localised, cases, the question is not who watered or didn't water a particular flower that dried in the garden, but what in general is the sort of entity that can save flowers from death by dryness. Seen from that perspective, interventionism is a theory that informs about means to an end.

However, if means–end relations are central, hypothetical interventions can be a problem. In my opinion one is again facing the problem of Frankenstein variables (Sect. 5.4). If someone asks, for instance, how to start a fire, interventionism can list matches, lighters, burning glasses or flints as potential means to an end. In addition to that, interventionism also lists hypothetical causes in the sense that they are impractical, such as flying combustible material into the sun. This is done—as mentioned in Sect. 2.4—in order to avoid anthropocentrism. On top of that, there might be still further causes that are hypothetical in the sense that no one knows whether they exist, such as the aliens I brought up as a hypothesis in Sect. 7.2. Where interventionism draws the line is unclear. Postulating aliens is speculative, but the variable A taking the values {*alien*, *no alien*} is well-defined. I don't see that there are any logical, conceptual or metaphysical reasons to reject that variable.

Then there is again the question of whether biological theory benefits from interventionist theory. The first thing to notice is that regulatory mechanisms, such as the *lac* and the *trp* operon, are feedback mechanisms. The usual representations produced by biologists contain loops—something that isn't allowed in the directed acyclic causal graphs used in interventionism. This, however, is only a minor inconvenience, since it is possible to construct directed acyclic causal graphs for feedback mechanisms, as shown by Marie Kaiser and Alexander Gebharter (2014). Whether those representations are useful for biologists is another question.

What biologists might be interested in is how they can reasonably exclude variables from their models. That, however, isn't done by interventionism with its liberal attitude towards hypothetical interventions. In that respect Hart and Honoré's approach (supplemented with Schurz's account of normality) and Hüttemann's theory have more to offer. The disposition-based process theory in particular makes it easy for biologists to justify why mere hypothetical causes are excluded: such candidate causes get excluded when they are no actual disturbances of the studied system. Even worse for interventionism is that it tends to overlook factual differences that are important to biologists. Although this is not directly related to the issue of omission, Waters (2007) is of course right to point out that interventionism fails to recognise the difference between actual and potential difference-makers in genetics (Sect. 2.6).

7.5 Counterfactual Theory

In this context, the significant feature of the counterfactual analysis (Sect. 2.2) is that David Lewis explicitly designed it as an account of token-*egalitarian* judgements (Sect. 2.2). This means that it is meant to distinguish between a broad notion of cause (*a* cause) and non-causes. In other words, Lewis analyses those factors or causes that are called 'conditions' by Hart and Honoré.

Under Lewis's analysis there is causal dependence between the gardener not watering the flower and the flower's death by dryness iff these two counterfactuals are true:

1. Had the gardener not watered the flowers they would have died.
2. Had the gardener watered the flowers they would have survived.

Possible world semantics tells one that these counterfactuals are true, since any possible world in which the gardener does not water the flowers and the flowers survive is less similar to actuality than any possible world in which the gardener does not water the flowers and the flowers die. Likewise, any possible world in which the gardener waters the flowers and the flowers survive is more similar to actuality than any possible world in which the gardener waters the flowers and the flowers die.

The problem is that these far-fetched statements are also rendered true:

1. Had the chancellor not watered the flowers they would have died.
2. Had the chancellor watered the flowers they would have survived.

Since so much hinges on the truth of these statements, I shall examine the crucial counterfactuals carefully. Is it really true that one cannot distinguish between the counterfactual 'had the gardener watered the flowers they would have survived' and the counterfactual 'had the chancellor watered the flowers they would have survived'?[7]

First I shall consider the actual world. I take into account the laws and the spatiotemporal distribution of facts. The actual world is a world in which the flowers die from dehydration and neither the gardener nor the chancellor waters the flowers. The laws are the laws of the actual world. Then I consider the counterfactual $A \Box\!\!\rightarrow C$, where A is the proposition 'the gardener does not water the flowers' and C is the proposition 'the flowers die from dehydration'. In natural terms, this means 'had the gardener watered the flowers (or had there not been the omission of the gardener not watering), the flowers would not have died from dehydration'.

For this counterfactual to be true it is required that any world in which both A and C are true is more similar to the actual world than a world in which A is true but C is not true. Now let us consider similarity between those worlds: There are various A-but-not-C worlds and various A-and-C worlds. There are many close A-and-C worlds that resemble actuality to a very high degree. The very similar worlds are all, by default, worlds that have the same initial conditions and the same laws. There is, for instance, a world in which everything happens as in the actual world except that there is a minuscule miracle such that a vortex on Jupiter slows down almost imperceptibly shortly before my flower dies. But is there any A-but-not-C world that exceeds such a degree of similarity with the actual world?

Given that exact similarities throughout large spatiotemporal regions have special weight, there are no such A-but-not-C worlds. The A-but-not-C worlds that exist are not similar enough. There are, for instance, worlds that share the laws and initial conditions of the actual world, except that shortly before my flower's dehydration some water miraculously appears so that my flower does not dehydrate. In such a case one has an extended similarity between spatiotemporal facts. However, the problem is that the required miracle is too exceptional. There are also A-but-not-C worlds in which no miracle occurs. But these worlds have other initial conditions, which means that there are great dissimilarities in the spatiotemporal distribution of facts. There is, for instance, an A-but-not-C world with the same laws but different initial conditions in which another person waters the flowers when the gardener has failed to water the flowers. But those worlds are ruled out by the default assumption that mismatches between the spatiotemporal distributions of particular facts are a heavy burden.

For the chancellor the reasoning is virtually the same: Let A be the proposition 'the chancellor does not water the flowers' and C be the proposition 'the flowers die from dehydration'. As before, the question is whether there are any A-but-not-C worlds that are more similar to actuality than A-and-C worlds. As before, the result must be that there are no such A-but-not-C worlds. There are again various close A-and-C worlds, but no A-but-not-C world can exceed them in similarity. As before, there are A-but-not-C worlds that closely resemble actuality with respect to

[7] The other two counterfactuals are automatically true, since they describe what happens in the actual world.

the spatiotemporal distribution of facts, except that they include a watering miracle so that C does not occur. And, as before, there are also A-but-not-C worlds with the same laws but different initial conditions where no miracles occur. Hence that counterfactual is also true.

A solution to the relevance problem would require rendering the counterfactual $A\Box\!\!\rightarrow C$ *true* in the case that the gardener does not water the flowers, but *false* in the case that the chancellor does not water the flowers. But one cannot show that $A\Box\!\!\rightarrow C$ is a false counterfactual unless one can show that there is some A-but-not-C world that is closer to actuality than any A-and-C world. This is impossible given the outlined concept of overall similarity between possible worlds.

However, it is not just a problem with this particular concept of similarity. Assume, for the sake of argument, that by default the integrity of laws outweighs the spatiotemporal distribution of facts. In that case one could argue that there is an A-but-not-C world with the same laws but different initial conditions in which the chancellor does not water the flowers but where someone else waters the flowers. However, such a modification of the concept of overall similarity would not help one to discriminate between the chancellor's not-watering the flowers and the gardener's not-watering the flowers either.

Given the modified concept of overall similarity, it would thus be reasonable to postulate that there is an A-but-not-C world with the same laws but different initial conditions in which the gardener does not water the flowers but someone else does. It is not just the Lewisian concept of overall similarity that poses a problem. The natural alternative to Lewis's suggestion of how to weigh the spatiotemporal distribution of facts against the integrity of laws runs into the same problem.

It is therefore not obvious how to tinker with the similarity concept. It would be easy to make a distinction between the gardener's not-watering and the chancellor's not-watering if it were admissible to give more weight to the spatiotemporal distribution of facts in one case and more weight to the integrity of laws in the other. I cannot see, however, how one could justify that double standard.

7.6 Causal Explanation as an Exit Strategy?

As it stands, counterfactual theory cannot counter the relevance problem. According to Lewis (2004a, pp. 99 ff.), this is, however, nothing to be worried about. Lewis's solution, also supported by Beebee (2004), is to distinguish causation from causal explanation. Why would that help? Briefly stated, their answer seems to be that non-entities, such as omissions, cannot be the source of a problem. They refrain from using Occam's razor to prune away at nothing, just as the Cyclopes turned away from their fellow after he had been blinded by Nobody.

The talk of absences is for Lewis a harmless fiction 'and we are within our linguistic rights to indulge in it' (Lewis, 2004a, p. 100). The only issue taken seriously by Lewis is the problem of the missing relatum, which he solves by claiming that, 'when the relata go missing, [the counterfactual analysis] can do without any causal relation at all' (Lewis, 2004b, p. 283). That there might be too many 'absences' is not a problem for Lewis or Beebee, since it isn't an ontological problem.

Granted, it might not be an ontological problem, but that doesn't mean that there isn't any problem remaining.[8] To be clear, the idea of differentiating between causation and causal explanation is perfectly reasonable. However, I don't think that Lewis's specific notion of causal explanation is convincing: the question of how to differentiate between relevant and irrelevant omissions is transformed into the question of how to differentiate between relevant and irrelevant causal explanations. Lewis's (2004a, pp. 99 ff.) answer is to discard the intuition that many explanations by omission are nonsense as secondary. According to Lewis, '*to explain an event is to provide some information about its causal history*' (Lewis, 1986, p. 217, italics in original). This is an incredibly low threshold.

To begin with, Lewis says that any information, even *incorrect information*, can be used to produce an explanation (Lewis, 1986, p. 218). His account is also tolerant of various sorts of information: there may be information that amounts to a certain *existential statement*, without further specification (Lewis, 1986, pp. 219 f.); it is possible that the information is an *analogy* with other familiar processes; and it is also possible to use *negative information*, i.e. 'information about what the causal history does *not* include' (Lewis, 1986, p. 220, italics in original). This list is not meant to be exhaustive. Lewis's central point is 'that we should be aware of the variety of explanatory information' (Lewis, 1986, p. 221). Ideally, an explanation would encompass all true information about an event's causal history. This would be called the '*whole* explanation' (Lewis, 1986, p. 219, italics in original). Actual explanations are, of course, never whole explanations: there are pieces of information about any event's past history that no one can know, either because the information is epistemically inaccessible or because it would be too much information for our limited cognitive capacity (Lewis, 1986, p. 219).

Lewis sketches the ideal of a whole explanation but says little about how to distinguish good from bad explanations. Following Lewis's account, all one can do is include some true detail—and if all true detail is included, i.e. if it is exhaustive, that is the whole explanation. While the criterion of truth should be fairly uncontroversial, the criterion of exhaustiveness is only intelligible against the backdrop of the Humean mosaic in which any detail can make a difference. This, however, is not how explanations in sciences, such as biology, are improved. Although good explanations will add more detail to precursor knowledge, this doesn't mean that all details matter in the expansion of an explanation in biology.

Take enzymes as an example. The function of enzymes depends on the structure of their active site. While the amino-acid sequence forming the active site tends to be highly conserved, other parts of the amino-acid sequence vary in a cross-

---

[8] Sarah McGrath (2005) also takes issue with the Lewis-Beebee position. The problem I have with McGrath's account is that I don't understand the notion of normality involved. It could be prototypical normality or, rather, an approximation to prototypical normality.

species comparison. The common explanation for this is that natural selection acts differently on those sequences. Alterations that change the active site are in the majority detrimental, i.e. they result in a loss of function. Natural selection punishes such changes and acts against them. If, however, a change occurs in a part of the enzyme that doesn't affect the structure of the active site, selection is neutral (Alberts et al., 2015, p. 15). Conserved sequences in a protein indicate a possible interaction with a ligand, while non-conserved sequences indicate that the specific arrangement of amino acids in that part of the molecule has no effect on the protein's function as an enzyme (Alberts et al., 2015, pp. 136 f.).

Although the non-conserved sequences are not part of the explanation of the function of the enzyme, they can reveal how species are related in the tree of life. This is possible because those differences in the amino-acid sequence are determined by DNA and therefore heritable (Alberts et al., 2015, pp. 15 f.).

Nature herself draws the line between the causally relevant and irrelevant properties of enzymes and she also decides which properties of enzymes are testimonies to the past and which are not. What about other structures, such as the cell, the tissue, the organism, the population, the ecosystem? Is it reasonable to suggest that any change anywhere is of causal significance? If one takes biological systems as a benchmark, the answer is: no.

Lewis's notion of a whole explanation is out of touch with the explanatory practice in biology. There is no reason to suggest that biologists should even aspire to map all the details of the cell, the organism or even the ecosystem—not because their research is interest driven, but because nature *teaches* them not to include every detail. Likewise, there is no reason to suggest that hypothetical non-interferences contribute to an ideal explanation.

Where to go from here? One could of course suggest supplementing the counterfactual theory with an account of causal explanation that respects explanatory practices in biology and other sciences. I would welcome such an effort. It is worrisome that Lewis didn't think of this as a valuable extension.

Beebee follows Lewis by arguing that omissions are not needed as truth makers of sentences. With that I agree. However, when Beebee alleges that those who want to distinguish relevant from irrelevant omissions are mistaking 'lack of explanatory salience for falsity' (Beebee, 2004, p. 307), her notion of explanation has to be questioned. If absences existed, one could reasonable argue that those strange absences are non-salient *because they are there*. If absences don't exist, one has to ask how 'absences can explain' implies '*all* absences are explanatory'. If Beebee's objective is to defend the position that the existence of omissions is no prerequisite for an account of causal explanation, there is no need for her to embrace the stronger thesis that all omissions are explanatory. She doesn't need to give in to the thesis that the non-interference of Godzilla is causally relevant to the world's history (Beebee, 2004, pp. 307 f.). Declining that thesis would in fact strengthen her argument, since she would avoid a difficulty that cannot be avoided by someone who accepts the existence of omissions. If the Godzilla-not-interfering omission exists, there is no escape from the conclusion that Godzilla not wreaking havoc is a cause of the way that world history has developed.

The fact that omissions are not needed as truth makers doesn't warrant the conclusion that the true counterfactual statement 'had the chancellor watered the flowers' is also a causal explanation. It is a true but *irrelevant* counterfactual statement. If one is talking about space aliens, it is a true but *absurd* counterfactual statement. Maintaining that these are also causal explanations requires an incredibly weak notion of explanation. If the gardener's employer asks the gardener why the flowers died, the gardener is expected to give an *explanation*. If the gardener lied and said that the neighbour poured salt over the flowers, this, although false, would still be an explanation. If the gardener said the chancellor or space aliens didn't intervene, this would not be an explanation by any commonly accepted standard.

One has to question whether, if a causal theory results in a myriad of irrelevant and absurd causal explanations, it can be considered a success. Bear in mind that those explanations concern *tokens*! If we were talking about types, as the interventionist theory would do, this would be less of an issue. The death of a particular flower at the edge of the Gobi Desert, although not caused by, is *causally explained* by the non-interference of the chancellor of the Federal Republic of Germany at the 23rd of March 2021 in the early hours of the afternoon local time. The same would be true of the president of the European Commission, the pope, the ship's cook on the infamous container vessel Ever Given, and so on. On a type-level account it can be admissible to claim that the interference of a head of government or the interference of a cook can avert the death of flowers, although one would still have to exclude the absurd cases. A token-level account doesn't afford these liberties. Hypothetical assumptions from out of the blue (even if they are true) do not explain anything. They are just counterfactual statements.

If proponents of the counterfactual theory still wanted to resist, there is one last argument I could think of for justifying that very weak notion of explanation: common sense doesn't care much about egalitarian token causation. Beebee is right when she says that salience is important. That is the central notion of cause (Sect. 2.6). Therefore intuitions about the notion of *a* cause may not be reliable. The counterfactual theory might show that Humeanism is tenable without a reasonable account of causal explanation, but only at the expense of decoupling itself from explanatory practices in the natural sciences and life in general. Why would one want to uphold such a position? It would be far more reasonable to agree that even the notion of *a* cause has boundaries and that the non-interference of Godzilla cannot be a causal explanation of any sort.

A common phrase in philosophical circles, also used by Beebee (2004, p. 300), is 'carving nature at its joints'. This phrase is usually thrown in when a theory presumably fails as an ontology. I suggest that this phrase is symptomatic of an unhealthy bias towards metaphysics instead of the physical world. The question should not be whether one succeeds in forcing nature into a metaphysical theory, but whether one succeeds in *reading nature's lips*.

7.7 Pluralism as an Answer?

The distinction between causation and causal explanation is no instance of causal pluralism in the technical sense. The particular Lewis-Beebee distinction is, however, structurally similar to Hall's position in the sense that it follows the same fault line, i.e. contiguous causation on the one side and something non-contiguous on the other side. This brings us back to the issues of Chap. 6. Does the pluralist position distinguishing production from dependence solve any problem a biologist might have with omissions?

According to Hall there are two concepts of cause that often coincide, although there are cases where only one of the two concepts applies. To distinguish the two concepts, it may be helpful to refer to the circumstances described in Sect. 6.1: The forest fire in June instead of May depends counterfactually on the rain in April, although the rain did not produce the forest fire.

By making that distinction Hall is certainly respecting the 'intuition of difference'. Similarly to the other positions discussed above, Hall gives up the idea of contiguity as far as dependency is concerned, arguing that causation in the sense of dependency can neither be local nor intrinsic (Hall, 2004, pp. 242 ff.). That Hall manages to uphold the contiguity thesis for causation in the sense of production might be seen as a benefit, although it doesn't solve any of the remaining issues concerning absences.

When it comes to the ontological status of omissions, Hall follows the same strategy as Lewis and Beebee, i.e. he argues that counterfactual statements can be true without omissions as truth makers (Hall, 2004, p. 254). Note that this argument is independent of the question of pluralism. Whether one distinguishes production from dependence doesn't have any consequences for the problem of the missing relatum.

The same is true for the relevance and absurdity problem. Hall does not even discuss the possibility of there being too many explanations, which is understandable since the productivity-dependence distinction is also a separate issue from there being irrelevant or absurd explanations. What is positive is that he is not actively arguing against differentiating between irrelevant and relevant explanations. This avoids tension between his position and the explanatory practices in biology. Since it is conceivable to supplement either the counterfactual theory or Hall's pluralist position with an account of causal explanation respecting biology and other sciences, there is nothing tipping the scales in either Lewis's or Hall's favour.

Next let me consider Glennan's position (Sect. 6.3). Glennan also distinguishes causal production from causal relevance. I suggested that Glennan's notion of canonical causation may justify viewing his overall account as an instance of arity pluralism, i.e. an account postulating concepts of cause and effect with different arity. I am putting this forward as a hypothesis that may be debated. Why is it interesting to consider Glennan's account from that perspective? Explanations resting solely on omission can be strange. Glennan's account clarifies in part why that is: they are incomplete.

Glennan's example of Mom not turning off the hose and thereby causing the basement to flood illustrates that point. One could say that the omission *Mom not turning off the hose* causes the basement to flood. Glennan argues, however, 'that Mom's failure to turn off the hose is not an event' (Glennan, 2009, p. 329). Although colloquially acceptable, one cannot literally say that the omission caused the basement to flood. What one would have to say is that the continuous flow of water caused the basement to flood by virtue of Mom not turning off the hose. The pattern of counterfactual dependence has to be embedded into the pattern of productive relations. However, Glennan notes that it is also true that productivity accounts cannot deal with omissions. Therefore, productive relations must also be embedded into a pattern of counterfactual dependence.

Applying the productivity concept is only meaningful against the background of the relevance concept and vice versa. They have to be integrated by the notion of a canonical causal judgement. This is to some extent similar to the way net and component effect are concepts that depend on each other, although the number of participating concepts is different (Sect. 3.3).

Glennan is silent on whether the chancellor, space aliens or other people are also causally relevant to the flooding in the basement. One might worry that Glennan's account is susceptible to such counterexamples, since it could be a fact that space aliens didn't turn off the hose. Yet it is also clear that the notion of a mechanism is important in Glennan's broader view of causation and causal explanation (Glennan, 1996, 2002, 2005, 2010b). This is where one can find a parallel to the disposition-based process theory. One can argue that mechanisms are in fact compound systems. The many parts of the mechanism are interacting subsystems. Similarly to Hüttemann, Glennan would have to exclude hypothetical disturbances. Whether this is as easy to do as in the disposition-based process theory, where it is done by restricting the account to actual disturbances, I cannot tell. It would seem that it is a fact about the *lac* operon that space aliens are not intervening. Since space aliens are not a component in that mechanism, one might want to argue that such facts have to be excluded. It could also be a concern that Mom, the hose and the basement are not a mechanism. What might solve this issue is that Glennan's notion of a mechanism is rather broad. He has argued that there are not only robust mechanisms, such as in biology, but also ephemeral mechanisms important for historical explanations (Glennan, 2010a).

Although I am sceptical that Glennan's account of canonical causal explanation contributes to the debate on frequency-dependent selection (Sect. 6.3), I will admit that this account sheds light on biological mechanisms. In contrast, Hall's account (despite its other merits) doesn't connect in any obvious ways to biological reasoning.

7.8 Conclusion

The theories discussed in this chapter are not equally suited to capturing biological reasoning about omissions. Normality approaches have a clear advantage over egalitarian theories. They deliver the answers needed by biology. In the case of the extended Hart and Honoré approach, the solutions grow quite naturally out of generalised evolutionary theory. Disposition-based process theory succeeds because biological entities are already conceived of as systems with dispositions by biologists, which means that no complicated re-description is needed.

Interventionism has a mixed outlook. It has some legitimacy as a type-level theory. The life sciences are not only interested in mechanisms but also in type-level questions such as 'does substance X cause disease Y?' However, biological explanation of such phenomena doesn't stop at the type level. After finding statistical evidence, researchers will usually seek a mechanist explanation of how a substance interacts with the organism on the molecular level. Note also that interventionism has no monopoly on such questions as these. Probabilistic theories (Sect. 2.3) can do that job, too. Where a unique answer might be expected, interventionism fails to deliver: how to exclude variables is not satisfactorily answered.

Counterfactual theory delivers—in principle—a reasonable answer to the question of why omissions are not detrimental to causal explanations. The attached theory of causal explanation is, nevertheless, disappointing. It isn't nuanced enough for biology. Of the two pluralist options presented by Hall and Glennan, the latter is the more interesting since there is an accompanying account of mechanism. This doesn't mean that I dismiss Hall's analysis. It is still a valuable analysis, but it is not the analysis needed in the context of this chapter.

What, then, are the conceptual distinctions that remain? I suggest choosing the disposition-based process theory, plus some type-level theory that sheds light on such questions as 'does substance X cause disease Y?' This latter approach need not be interventionism, as there are other options. However, further theories are not in my opinion required for the issue of omissions in biology. Although I advocate a little pluralism, I am not recommending an endless proliferation of concepts—but I am also not ruling out the possibility that *other* problems (of which I am unaware at this point) may present reasons to bring in other theories.

References

Alberts, B., Johnson, A., Lewis, J., Morgan, D., Raff, M., Roberts, K., & Walter, P. (2015). *Molecular biology of the cell* (6th ed.). New York and London: W. W. Norton & Company.

Beebee, H. (2004). Causing and nothingness. In J. Collins, N. Hall, & L. A. Paul (Eds.), *Causation and counterfactuals* (pp. 291–308). Cambridge, MA and London, England: MIT Press.

Campbell, N. A., & Reece, J. B. (2005). *Biology* (7th ed.). San Francisco: Pearson Benjamin Cummings.

Dowe, P. (2000). *Physical causation*. Cambridge: Cambridge University Press.

Dowe, P. (2001). A counterfactual theory of prevention and 'causation' by omission. *Australasian Journal of Philosophy, 79*(2), 216–226. https://doi.org/10.1080/713659223

Fair, D. (1979). Causation and the flow of energy. *Erkenntnis, 14*(3), 219–250. https://doi.org/10.1007/BF00174894

Gebharter, A., & Kaiser, M. I. (2014). Causal graphs and biological mechanisms. In M. Kaiser, O. Scholz, D. Plenge, & A. Hüttemann (Eds.), *Explanation in the special sciences* (Vol. 367, pp. 55–85). Dordrecht: Springer. https://doi.org/10.1007/978-94-007-7563-3_3

Glennan, S. (1996). Mechanisms and the nature of causation. *Erkenntnis, 44*(1), 49–71. https://doi.org/10.1007/BF00172853

Glennan, S. (2002). Rethinking mechanistic explanation. *Philosophy of Science, s69*, S342–S353. https://doi.org/10.1086/341857

Glennan, S. (2005). Modeling mechanisms. *Studies in History and Philosophy of Biological and Biomedical Sciences. Special Issue: Mechanisms in Biology, 36*(2), 443–464. https://doi.org/10.1016/j.shpsc.2005.03.011

Glennan, S. (2009). Productivity, relevance and natural selection. *Biology and Philosophy, 24*, 325–339. https://doi.org/10.1007/s10539-008-9137-7

Glennan, S. (2010a). Ephemeral mechanisms and historical explanation. *Erkenntnis, 72*(2), 251–266. https://doi.org/10.1007/s10670-009-9203-9

Glennan, S. (2010b). Mechanisms, causes, and the layered model of the world. *Philosophy and Phenomenological Research, 81*(2), 362–381. https://doi.org/10.1111/j.1933-1592.2010.00375.x

Hall, N. (2004). Two concepts of causation. In J. Collins, N. Hall, & L. A. Paul (Eds.), *Causation and counterfactuals* (pp. 225–276). Cambridge, MA and London, England: MIT Press.

Hart, H. A. L., & Honoré, T. (1985). *Causation in the law* (2nd ed.). Oxford: Clarendon Press. (Original work published 1959)

Hume, D. (2000). *A treatise of human nature.* In D. F. Norton & M. J. Norton (Eds.). Oxford, New York: Oxford University Press. (Critical texts of the first editions of *A treatise of human nature*, books 1–3, and *An abstract of… a treatise of Human Nature.*)

Hüttemann, A. (2013). A disposition-based process-theory of causation. In S. Mumford & M. Tugby (Eds.), *Metaphysics and science* (pp. 101–122). Oxford: Oxford University Press.

Lange, M. (2002). *An introduction to the philosophy of physics. Locality, fields, energy, and mass.* Oxford, Malden: Blackwell Publishers.

Lewis, D. (1986). *Philosophical papers* (Vol. 2). New York, Oxford: Oxford University Press.

Lewis, D. (2001). *Counterfactuals.* Malden, MA: Blackwell Publishers. (Original work published 1973)

Lewis, D. (2004a). Causation as influence. In J. Collins, N. Hall, & L. A. Paul (Eds.), *Causation and counterfactuals* (pp. 75–106). Cambridge, MA and London, England: MIT Press. (Reprinted from *Journal of Philosophy, 97*(4), 182–197 (2000); minor revisions have been made for consistency)

Lewis, D. (2004b). Void and object. In J. Collins, N. Hall, & L. A. Paul (Eds.), *Causation and counterfactuals* (pp. 277–290). Cambridge, MA and London, England: MIT Press.

Matthen, M., & Ariew, A. (2009). Selection and causation. *Philosophy of Science, 76*(2), 201–224. https://doi.org/10.1086/648102

McGrath, S. (2005). Causation by omission: a dilemma. *Philosophical Studies, 123*(1–2), 125–148. https://doi.org/10.1007/s11098-004-5216-z

Moore, M. S. (2009). *Causation and responsibility. An essay in law, morals, and metaphysics.* Oxford: Oxford University Press.

Schaffer, J. (2000). Causation by disconnection. *Philosophy of Science, 67*(2), 285–300. https://doi.org/10.1086/392776

Schaffer, J. (2004). Causes need not be physically connected to their effects: The case for negative causation. In C. Hitchcock (Ed.), *Contemporary debates in philosophy of science* (pp. 197–216). Malden, MA: Blackwell Publishers.

Schurz, G. (2001). What is 'normal'? An evolution-theoretic foundation for normic laws and their relation to statistical normality. *Philosophy of Science*, *68*(4), 476–497. https://doi.org/10.1086/392938

Stapleton, J. (1994). *Product liability*. London: Butterworths.

Waters, C. K. (2007). Causes that make a difference. *The Journal of Philosophy*, *104*(11), 551–579. https://doi.org/10.5840/jphil2007104111

Woodward, J. (2003). *Making things happen. A theory of causal explanation*. New York: Oxford University Press.

Chapter 8
Epilogue

Abstract This chapter summarises the book's overall argument that causal pluralism doesn't deliver on its promises. I also provide critical comment on the recent suggestion to speak of 'reciprocal causation' and I suggest that researchers should be more open about the fact that they do (and in my opinion also need to) compare and weigh causal contributions against each other. The chapter closes with a warning not to overestimate the scope for reasonable applications of causal pluralism.

Keywords Causal monism · Causal pluralism · Conceptual plurality · Cyclical models of causation · Dominant cause · Reciprocal causation

Where are the limits of conceptual pluralism? They are probably not as broad as one might expect if one took the enthusiasm for pluralism in recent times as a reference point.

The motivation of conceptual distinctions is not always the same and not every choice of concepts is legitimate. Part of the motivation for conceptual plurality may simply be the conviction that there are no universal answers and that one therefore has to be flexible with one's causal concepts. But that isn't all. Causal concepts also get generated when an effort is being made to present certain views in a more favourable light.

Who would pay attention to someone who said that all those discoveries about the structure of DNA and the molecular structure of the cell might well be very impressive, but what about those other causal factors in the evolutionary past which brought about the traits of organisms living in the present? Just imagine how such a message would be received if it were rounded off by saying, 'And by the way, all those causal relations are no different in kind from the causal relations with which you are already familiar.' That appearances mattered more than reality in Mayr's account of ultimate and proximate causes is a criticism that has been put forward before. What critics have failed to do, however, is show why someone who takes Mayr's position has to give up that view in favour of another position. In the absence of any mutually exclusive alternatives, Mayr's position might live on indefinitely.

K. Ehrenstein, *Causal Pluralism in the Life Sciences*,
History, Philosophy and Theory of the Life Sciences 25,
https://doi.org/10.1007/978-3-030-87942-6_8

In such a situation an account of conceptual causal pluralism can be useful, since it brings into sharper focus how some distinctions of causal concepts are trivial. One may argue that this is often obvious anyway. I suggest, however, that one might hoist Mayr by his own petard and put a label on such trivial distinctions by calling them 'instances of monism'.

The initial challenge was of course to disentangle Mayr's partially contradictory position until it could clearly be seen how his position exemplifies monism. I dare say that incoherence itself might even be seen as one of Mayr's defensive strategies. By being vague, Mayr constantly forces critics to fill in the gaps of his account. If, as happened in his dispute with Francis (Sect. 4.6), a critic should articulate what seems to be implicit in the ultimate-proximate distinction, Mayr blithely deflects this criticism by claiming that the critic has somehow overlooked the fact that Mayr now refers to ultimate causes as 'evolutionary causes'.[1]

Although any topic in biology can be quite complex, this should not be taken as an excuse for shying away from clearly articulating what ultimate and proximate causes are supposed to be. The only clear conception of what those terms mean can be found in the earlier sources from which Mayr got the original idea (Sect. 4.1). But even then the ultimate-proximate account appears to be nothing more than explanatory relativity in disguise, which is an important reason for dismissing the account (Sect. 4.8).

Does this mean that all conceptual distinctions of causes in biology have to be useless? I don't think so. Other conceptual distinctions, even if one has to label them as 'monism', can be legitimate *as long as they are not sold as referring to some special sort of causation*. One should not, however, follow Mayr's example of coining vague causal terms. I am, therefore, critical of some of the terminology circulating among critics of Mayr, such as 'reciprocal causation' and the idea of cyclical models of causation (Laland et al., 2013). It is evident that 'cyclical' cannot be taken literally and 'reciprocal' is nothing more than the idea that there are temporally alternating causal influences of different causal factors. I am worried that this is rhetorically inflated language.

The problem is not just that this is obscure language, but also that some arguments trivially collapse because of this mental picture of cycles. If any viewpoint can be taken, it doesn't matter how many backwards-pointing arrows are added to a diagram illustrating the evolution of the peacock's tail (Sect. 4.8). Cyclical diagrams, although a useful means of communication in other contexts, are *no* compelling response to Mayr's dogma of ultimate causes being in charge of proximate causes. What critics of Mayr should emphasise in their arguments is that, depending on the *concrete* trait under investigation, a causal factor sometimes ranks as a condition and sometimes as a dominant cause.

---

[1] In the earlier version of my manuscript, I was critical of Mayr but had not yet settled on quite such a strong verdict. I tried very hard to develop some sort of reasonable and consistent interpretation. My view of Mayr hardened, however, with the help of an anonymous reviewer who pointed out even more problems of which I hadn't been aware.

More could be achieved if biologists were unambiguously open about the fact that they are often arguing about the dominant causes of a specific trait—such as the peacock's tail. Here sexual selection appears to be the right answer if one is asking for the dominant causal factor. In other cases such as convergent evolution of body shapes of aquatic animals, natural selection is likely to be the dominant cause. In yet other cases a trait might be owed to the fact that developmental pathways act as a constraint. One may also expect that it can be unclear whether there is a dominant cause. It is at any rate reasonable to assume that it will not just be a simple matter of finding the dominant cause, but will also require an estimation of how different influences (such as being attractive to peahens and not being caught by a predator) trade off against each other.

When sometimes one group of factors is given more weight than another group, it indicates that biologists are already following the intuitive idea that causes and conditions are not the same. Confusion arises when one relapses into the naive philosophical view of thinking that any distinction between causes and conditions has to be interest-laden.

Although interest can be a confounding factor, it is not in all cases the driving factor determining the context. For instance, the fact that natural selection punishes some deviations so harshly points to the significance of developmental biology. If developmental pathways are delicate in the sense that they tolerate only very limited changes without collapsing into chaos, then the endless supply of variation envisaged by Mayr is not there. In such a scenario, selection becomes a condition instead of a dominant cause (Sect. 4.8). And there is of course also the question of whether variation, as it occurs in a particular case, is due exclusively to the genetic underpinning (Sect. 5.6). None of these questions is adequately dealt with by insisting that there are many perspectives and that explanatory relativity allows one to select whichever factor one likes.

That biologists are often interested in the asymmetry of causal contributions is also evident when one considers the permissive-instructive distinction. Here, too, the question is whether there are causal factors, such as the sequence of the DNA determining the sequence of the RNA, occupying a prominent role.

One may of course ask how all these ideas of causal asymmetry are related. It is evident that there is not just one strategy for determining causal asymmetries, but many, including the idea of influence or specificity, the idea of prototypical and statistical normality, the actual difference-making account and the disposition-based process theory.

It is also interesting that the emphatically egalitarian counterfactual analysis of causation has changed its character over time. Lewis's original theory aimed at a non-discriminatory notion of cause (Sect. 2.2). That the revised account talks about influence is remarkable against that background. The way Lewis used influence as a tool in his later theory suggests that he still intended to draw a line between causes and non-causes instead of causes and conditions. However, by his own admission, '[i]nfluence admits of degree in a rough and multidimensional way' (Lewis, 2004, p. 92). Why would one rank causes according to their influence if one's initial project was to analyse an egalitarian notion of cause? Is it just a tool for drawing a fuzzy boundary between non-causes and causes? Or, rather, a tool for identifying one or more causes dominating other causal factors?

How all these ideas of causal asymmetries fit together I cannot tell. What I can say is that the philosophical reflex to discard as interest-driven any attempt at giving more weight to some causal factors than to others is short sighted. If there is anything that philosophical theory can give to biological theory, it is a principled account of how to weigh causal factors against each other. In that respect, interventionist theory makes an interesting contribution by quantifying causal specificity (although such quantification is of course not exclusively bound to that theory).

However, an issue in which interventionism doesn't excel is the analysis of concrete molecular mechanisms. The interventionist account doesn't advise how to choose reasonable variables, and the way that Woodward fails in his analysis of protein biosynthesis vividly demonstrates this shortcoming of the account.

At this point, one may ask how much pluralism is actually needed. Do we need more than one theory at our disposal? I would suggest that interventionism, and similar theories conceiving of causation as relations between variables, are powerful heuristic methods even though they are not suited as ontic accounts. Before a phenomenon is operationalised in terms of variables it needs to be analysed by an account that has a better grasp of physical properties. My discussion of protein biosynthesis in Sect. 5.2 shows that such an analysis can proceed without a concrete theory of causation. Yet it is also evident that among the theories presently available the disposition-based process theory is best suited to the task. Biology describes many phenomena in terms of systems, even down to the molecular level where proteins and other molecules can be conceived of as systems manifesting their dispositions in different degrees. However, since disposition-based process theory doesn't tell us directly how much influence a molecule such as DNA exerts over a mechanism (which can be conceived of as an arrangement of interacting systems forming a compound system), other theories such as interventionism can act as a useful completion.

Less urgent appears to be the export to biological theory of other philosophical ideas such as distinguishing production from dependence (Chap. 6). My analysis shows that there is even quite a bit of variation between different versions of that distinction. Yet none of these accounts is in any obvious way relevant for biology. Note that one cannot conclude that these accounts are irrelevant in all circumstances. There may be other currently unknown applications. So far, however, these accounts appear to be more or less self-referential. Hall's analysis is a challenge to counterfactual theory, but perhaps also an opportunity for asking whether there is more than one notion of locality. Hüttemann (2013), who claims to solve Hall's problem within his disposition-based process theory, appears to presuppose a different notion of locality from that put forward by Hall (2004). I don't want to digress too much, but Hall's notion of locality appears to be equivalent to contiguity, while Hüttemann's notion of locality is explicated as spatiotemporal proximity.

Such comparisons between theories can be illuminating in a broader philosophical context, so I am not arguing that philosophers should stop thinking about these distinctions. My point is that what is *so far* known to motivate these proposals cannot demonstrate how these distinctions solve any problem in biological theory.

However, the limits of conceptual causal pluralism have to be narrower than the broad field of philosophical speculation. Where causal concepts are applied instead of being scrutinised as philosophical ideas, one has to insist that only concepts that matter be brought into debates in other fields. Otherwise philosophy becomes a source of cheap diversions for pushing through arbitrary agendas.

An important lesson from the analyses conducted in this book is that *few problems actually hinge on the question of whether one subscribes to monism or to pluralism*.

The dispute between Mayr and his critics eventually comes down to the question of how to weigh various causal factors against each other. What names one gives to certain groups of causal factors is irrelevant. It is even conceivable that researchers could have had such a dispute without the terminology of 'ultimate' and 'proximate causes'. Without that terminology the actual discussion may have been much clearer.

Other distinctions may look as if some sort of 'plurality' is at work. The terminological distinction between permissive and instructive causes may appear to be conceptual pluralism. Yet as soon as one recognises that these terms translate into 'low' and 'high influence' or 'specificity' one has to conclude that they are merely terms that have their place in monist theories such as counterfactual or interventionist theory (if one ignores the dimension of arity pluralism). Although these are, in contrast to 'ultimate' and 'proximate cause', legitimate concepts, they are not indicative of any meaningful conceptual pluralism.

Arguments about phylogenetic inference, genetic drift or frequency-dependent selection are also unaffected by the question of whether one subscribes to a proper conceptual pluralism of the sort advocated by Sober or Glennan.

Causation by disconnection, which could invite the view that one may have to distinguish production from dependence or quasi-causation, is also no justification for pluralism. Sorting out bogus omissions isn't done by conceptual distinctions.

The evidence 'against' pluralism in this book is of course only circumstantial and inductive, in the sense that my conclusion is probable without being necessary. Yet it is no less conclusive than the evidence that is sometimes brought up against monism (Sect. 1.5). A common strategy for arguing against monism is to point out counterexamples that demonstrate its shortcomings. My strategy in this book has been no different. I have simply raised awareness of the shortcomings of pluralism.

However, my aim has not been to *disprove* pluralism. That is as impossible as proving pluralism. The more important consideration is that it is sometimes of little relevance whether pluralism is true or not. One should instead ask whether *specific* problems can be countered with a pluralist strategy. In the end I can only voice my scepticism towards the contention that pluralism is always advisable, since it can also be a source of harmful confusion. One should be aware of pluralism as an option, without being blinded by its promises.

References

Hall, N. (2004). Two concepts of causation. In J. Collins, N. Hall, & L. A. Paul (Eds.), *Causation and counterfactuals* (pp. 225–276). Cambridge, MA and London, England: MIT Press.

Hüttemann, A. (2013). A disposition-based process-theory of causation. In S. Mumford & M. Tugby (Eds.), *Metaphysics and science* (pp. 101–122). Oxford: Oxford University Press.

Laland, K., Odling-Smee, J., Hoppitt, W., & Uller, T. (2013). More on how and why: Cause and effect in biology revisited. *Biology and Philosophy*, *28*(5), 719–745. https://doi.org/10.1007/s10539-012-9335-1

Lewis, D. (2004). Causation as influence. In J. Collins, N. Hall, & L. A. Paul (Eds.), *Causation and counterfactuals* (pp. 75–106). Cambridge, MA and London, England: MIT Press. (Reprinted from *Journal of Philosophy*, *97*(4), 182–197 (2000), minor revisions have been made for consistency)

Index

K. Ehrenstein, *Causal Pluralism in the Life Sciences*, History, Philosophy and Theory of the Life Sciences 25,
https://doi.org/10.1007/978-3-030-87942-6

The manufacturer's authorised representative in the EU is Springer Nature Customer Service Centre GmbH, Europaplatz 3, 69115 Heidelberg, Germany. If you have any concerns regarding our products, please contact ProductSafety@springernature.com

Printed and bound by CPI Group (UK) Ltd, Croydon, CR0 4YY
22/09/2026
02220982-0005